Caribbean Sea Slugs

A field guide to the opisthobranch mollusks from the tropical northwestern Atlantic

By

Ángel Valdés
Jeff Hamann
David W. Behrens
Anne DuPont

Sea Challengers Natural History Books, Etc.
Gig Harbor, Washington 98335

2006

A SEA CHALLENGERS NATURAL HISTORY BOOKS PUBLICATION

First Edition
First Printing

FRONT COVER

Hypselodoris bayeri, Biscayne Bay, Florida by Jeff Hamann

First Published in 2006 by
Sea Challengers Natural History Books, Etc.
5091 Debbie Court
Gig Harbor, Washington 98335
United States of America

ISBN 0-9700574-2-3

Printed in Hong Kong through Global Interprint Inc., Santa Rosa, California 95401

Format, Cover Design and Prepress by Ángel Valdés

Most of the field work and photography for this book were done from 1983-2000 by Jeff Hamann on the Gloriamaris, a 50-foot Prout Catamaran. The Gloriamaris was purchased by the Hamann Family in England and delivered to Ft. Lauderdale on Mom's birthday, May 5th, 1983. I, Jeff, am the oldest of Bob and Dot's four boys. Over the subsequent 17 years we covered nearly all of the Caribbean Islands as well as the coasts of Venezuela, Columbia, Panama, Nicaragua, Honduras, Guatemala, Belize and Mexico.

We kept a caretaker/captain on board and took turns moving it and enjoying different places, sometimes sailing and sometimes motoring with the two 25 hp diesel motors. Over the years, we caught marlin and dorado in St. Vincent, swam with whale sharks in Belize, explored the wreck of a Spanish galleon in Cuba, and went through the Panama Canal. But we also had our captain arrested in Roatan, weathered a hurricane in Culebra, had the boat confiscated in Puerto Rico, and were boarded and robbed by two Garifanos in Honduras. It has been an incredible adventure and the best parts were the diving and the branching.

Each year I made two to five field trips to the boat as it made its way from place to place. Other opisthobranch aficionados like Hans Bertsch, Terry Gosliner, Sandra Millen, Wes Farmer and Dave Behrens all found themselves on the Gloriamaris at different points along the way. The Gloriamaris proved to be a great dive boat and floating lab. The catamaran hull configuration made the boat generally stable enough for lab photography and the ample deck space allowed for a dive compressor and six scuba tanks. A folding swim platform extended 12 feet from hull to hull. A Carib RIB dinghy with a 25 hp outboard was the final piece of equipment that allowed us to explore and dive up to a few miles from the Gloriamaris.

Over one third of the specimens recorded were found at night. Rubble bottoms in current ravaged channels were particularly productive. I would often clip the dinghy anchor line to my BC and simply drag it along with us on a night dive but sometimes it would drag me along with the wind and current. Other times we would dangle a strobe light from the swim step and try to navigate a rough circle route back to the boat. We rarely missed a night dive while we were on the boat and we occasionally even made two dives in one night.

Not only did I take over 20,000 photographs of nearly 250 species of opisthobranchs on the Gloriamaris, but I also watched my three girls grow from infants to nudibranch savvy snorkelers and then into young ladies. The year we bought the Gloriamaris, Tisha was three and my wife, Marsha, gave birth to Elita. Dana came three years later. As toddlers they loved to run around the deck sans diaper and play on the beach. In the elementary years they marveled at all of the sea life we were blessed to observe. They would try and wait up at night knowing that Dad would bring back something strange to look at. And then in the teen years, friends, fishing and wakeboarding were important interests. The Gloriamaris is now in the Pacific and the Girls are all grown up.

PREFACE

Crystal clear waters, white sand beaches, pristine reefs and Jimmy Buffet songs are what most people think of when they imagine the waters of the Caribbean. These are not what those of who are fascinated with nudibranchs think about when we think of the Caribbean as habitats for finding the great diversity of species in the Western Atlantic. Nudibranch enthusiasts would more likely be found turning rubble off the Burger King at Grand Cayman than swimming with the rays at Sting Ray City or diving the deep dropoffs. And while we may be thinking about a cheeseburger in paradise after the hunt, we are happiest snorkeling around the mangroves or carefully searching blades of turtle grass. These specialized habitats are the haunts of the approximately 300 species of nudibranchs that make up the rich biodiversity of the Caribbean.

The Caribbean is known to most enthusiasts of tropical marine life from Paul Humann's and Ned DeLoach's pioneering field guides that provide a wonderful taste of the diversity that inhabits this tropical region that borders our southeastern shores. Paul and Ned's discerning eyes have only scratched the surface of what lies beneath the waters of the Caribbean in the nudibranch realm. Thanks to the decades of exploration by Jeff Hamann, and the subsequent studies by Anne DuPont, Dave Behrens and Ángel Valdés, a vast diversity of stunning images provides the unique opportunity to produce the first comprehensive guide to the opisthobranch fauna of the Caribbean. All the species known only from preserved animals described by Rudolf Bergh and Ernst and Eveline Marcus are now brought to life by this comprehensive compilation. All of the many species described by Jesús Ortea are now brought together for the first time. This work is truly a feast for the eyes, with amazing patterns of warning coloration that rival the most colorful molas produced by the Cuna Indians of Panama.

Not only does this book provide stunning photos, it also breaks new ground in demonstrating the variation within species. It also provides a great amount of information about the basic biology of opisthobranchs, including food preferences of the highly specialized diet of most nudibranchs and their relatives, their egg masses and defensive mechanisms. The individual species treatments add a great deal of knowledge about the taxonomy and distribution of species and often include scanning electron micrographs of the radula of many species that expands the utility of this work from spanning the dive enthusiast to the professional malacologist.

The Caribbean, like many of the world's oceans, is facing a crisis. Reefs are bleaching or are dying as we humans induce global climate change through our own activities. It is absolutely necessary to have comprehensive works such as *Caribbean Sea Slugs* to document the biodiversity of the world and how much is at stake. Hopefully, this book will inspire all of us to find solutions to help preserve the diversity that hangs in the balance.

Terrence M. Gosliner, Ph.D.
Senior Curator
California Academy of Sciences

ACKNOWLEDGEMENTS

Several professional and amateur photographers, divers and scientists contributed photographs, specimens and/or other information; they are (in alphabetical order): Néstor Ardila, Patricia Armes, Andrea Bernecker, Hans Bertsch, Joyce and Frank Burek, Kerry Clark (deceased), Yolanda Camacho, Carol Cox, Malcolm Edmunds, Richard Fox, Bill Frank, Terry Gosliner, Larry Harris, Leslie Harris, Jamie Holdorf, Stephane Hourdez, Linda Ianniello, Rebecca Jackrel, Kevin Johnson, Sally D. Kaicher (deceased), Ernst Marcus (deceased), Kaniaulono Meyer, Leopoldo Moro Abad, Betsey Putnam Hansen, Colin Redfern, Lazara Ruda, Edward Ruppert, Herb Segars, Judy Townsend, and Jack Worsfold.

Malcolm Edmunds, Terry Gosliner, Kathe Jensen, Patrick Krug, and Sandra Millen provided constructive comments on the manuscript. Elita Hamann edited the manuscript for grammar and consistency.

Two websites, the Sea Slug Forum (www.seaslugforum.net) and Malacolog - Western Atlantic Mollusk Database (data.acnatsci.org/wasp) provided useful information for identifications, synonymies, and geographic ranges of species.

The US National Science Foundation, through the grants DEB-9978155 (to Gosliner & Valdés), DEB-0329054 (to Gosliner & Valdés), and DBI-0216506 (to Valdés *et al.*), supported several research projects, field expeditions, and scanning electron microscope work that produced original information and materials used in this book.

INTRODUCTION

The Caribbean is visited by more divers than any other oceanic region on earth. It is impossible to estimate how many of us enter the water each day at the hundreds of diving sites on islands and the continent in search of breathtaking underwater experiences.

Over the past several years one group of small marine animals, the opisthobranchs, also commonly known as sea slugs, and particularly a group of them called nudibranchs, has gained increasing attention and popularity. Dozens of books have been published around the world to identify opisthobranch species, with the notable exception of the Caribbean, for which there are no comprehensive guides available. The closest the diver has been able to come to a Caribbean sea slug reference book are Paul Humann & Ned DeLoach's (2002) – *Reef Creature Identification* (2nd Edition) and Colin Redfern's (2001) – *Bahamian Seashells*. While both are excellent illustrated guides for invertebrates and mollusks respectively, their coverage of sea slugs is limited.

The scientific literature on Caribbean opisthobranchs is rich. The most substantial contributors to the knowledge of the Caribbean sea slugs have been Ernst and Eveline Marcus, a German couple who immigrated to Brazil escaping Jewish persecution in Europe. Altogether the Marcus' described about 180 species in the tropical western Atlantic between 1956 and 1977. Most of their papers were based on specimens collected by other persons and sent to the Marcus' already preserved. Because of this, the external coloration was not described for

Map of the geographical coverage of this guide (A. Valdés)

most species and it is now very difficult to recognize many of them based on these papers. Other significant contributors to the scientific literature were Kerry Clark (USA) who worked on Bermudan and Floridan species, particularly sacoglossans, T. E. Thompson (UK), who published on Bermudan and Jamaican sea slugs, and Malcolm Edmunds (UK) who worked on species from Jamaica and Barbados. In recent years Jesús Ortea (Spain) and his team have been very productive describing a number of species from the area.

Geography, biodiversity and biogeography

The scope of this book is the tropical north-western Atlantic, defined here as the warm body of water including the Atlantic coast of Georgia and Florida, the Gulf of Mexico, the eastern coast of Central America, the northern coast of South America, and all of the offshore islands from Bermuda and the Bahamas to the Greater and Lesser Antilles. This book is restricted to shallow water species, defined as those found in SCUBA diving depths, down to 80 meters (about 260 feet). The Caribbean deep-sea remains largely unexplored, although a few species of sea slugs have been described.

Aside from the nearly 1,000 islands, cays, and atolls distributed across this area, one very significant feature found along the coast of Mexico, Belize and Honduras is the second longest barrier reef in the world, and the longest one in the northern hemisphere. In fact, the Caribbean contains the only substantial coral reef formations in the Atlantic basin.

The Caribbean is known for its current swept, steep walls, spur and groove reefs, patch reefs, turtle grass beds, sheltered lagoons and mangroves. Reef formations provide the habitat for gardens of huge colorful sponges and gorgonian soft corals. Turtle grass beds present a completely different flora and fauna, while the complex root systems of the mangrove forests create a reef-like maze providing sanctuary for thousands of species of fish and invertebrates.

Turtle grass beds have the ability to greatly modify the physical, chemical, and geological properties of coastal areas. They provide nutrients, primary energy, and habitats to sustain coastal fishery resources; create foraging grounds for endangered species; and sustain biological diversity and productivity. And, they are full of sea slugs!

The Caribbean is peppered with thousands of exotic cays (photos D. Behrens, left; © B. Jones & M. Shimlock, right)

Mangroves are one of the most highly productive habitats in the Caribbean. While the Caribbean has only 4 of the 80 mangrove tree species found worldwide they are no less productive and unique. Like coral reefs and seagrass beds, the environmental values of mangrove forests are critical to fisheries, wildlife resources, and biodiversity. The matrix created by

Turtle grass beds provide a diverse habitat and substrate for many smaller species (photos D. Behrens)

their tangled and branching root systems provide infinite real estate for encrusting animals such as corals and sponges, as well as the perfect refuge for juvenile fishes and motile species like lobsters and shrimp. In this habitat also, sea slugs abound.

We are often told by fellow divers that they never see sea slugs in the Caribbean, like they do in the tropical Indo-Pacific. While the number of species found in the two corners of the globe differs dramatically, you will be surprised to know that the Caribbean is home to over

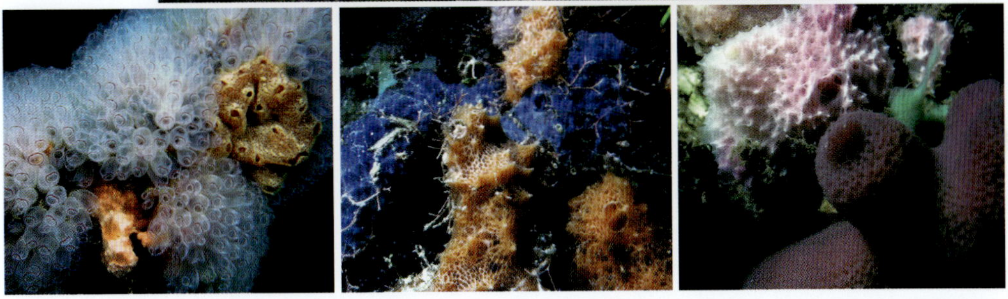

The root system of mangrove forests provide habitat for hundreds of encrusting species like these sponges and tunicates and is the perfect place to search for nudibranchs (photos D. Behrens)

Coral head habitat like this is not the best place for nudibranch hunting (photo © B. Jones & M. Shimlock)

300 species, representing every major lineage found in the tropical Pacific. However, Caribbean species have less abundant populations than those in the tropical Indo-Pacific, and their habits are more cryptic, making finding sea slugs much more challenging and time consuming, for the diver or tidepooler.

Where to Look

While many species of sea slugs can be easily found on the reef by searching carefully, many species are more common in areas and habitats not frequented by recreational divers, especially those diving commercial dive boats. We hope to introduce you other places to dive to broaden your Caribbean dive and snorkeling experience, and help you learn more about these fascinating and very colorful marine creatures.

If you really want to expand your life list of opisthobranchs or study these animals in more detail, you need to know where to find them. Here are a few general tips. You will find more specifics in the species descriptions chapters.

Believe it or not, live coral reefs are rarely productive for sea slugs. While you will find many species there, you are better off looking elsewhere. Channels with a lot of current and rubble bottom are a good bet, although you will have to roll the rubble over to look underneath, or dive at night. Look closely at hydroids on sea weed or walls. *Sargassum* algae is another hangout for several species. Some of the most productive habitats are underneath boat docks, cave roofs, mangroves, and mooring chains. These substrates provide lots of underwater real estate for the prey species of opisthobranchs to live on. Find their food and you have improved your chances of finding slugs.

Floating alga is a hangout for several species (photo A. DuPont)

Sponges (left), hydroids (center) and anemones (right) are some of opisthobranchs' favorite foods (photos D. Behrens)

Most dorid nudibranchs feed on sponges that grow on the undersides of dead coral rubble and boulders. Generally, aeolid nudibranchs feed on hydroids and other cnidarian species such as sea anemones. The sacoglossans or sap-sucking slugs feed on algae, so close inspection of the green tufts of *Penicillus* and *Caulerpa* will prove rewarding. Sea hares are also algae eaters, so you will find them in the turtle grass beds. Head shield slugs like *Gastropteron* and *Chelidonura* are generally found in calm sandy areas from 2 to 10 meters deep.

Clumps of algae are where sacoglossans or sapsucking slugs are found (photos D. Behrens)

Lastly, if you see egg ribbons, coils or masses on a substrate stop and look more closely. The chances are good that the adult that laid them is nearby.

Egg ribbons are a dead giveaway that an adult may be nearby (photos J. Hamann)

Just remember, in most countries, permits are necessary to collect, export and/or import live or dead sea slugs or other marine invertebrates. Check with the local authorities before engaging in any collecting or transporting of opisthobranchs.

Morphological diversity in opisthobranchs (photos J. Hamann)

Opisthobranch diversity and evolution

Sea slugs are the most structurally diverse group of mollusks, showing an evolutionary trend towards the reduction, internalization and independent loss of the shell. The loss of the protective shell has forced these organisms to depend upon alternative defensive systems. The most common defenses used by sea slugs are chemical deterrents and repellents. These chemicals are either obtained from the prey species they feed on or synthesized within the slug itself.

Defensive secretions in sea slugs (photos J. Hamann)

In many cases chemical defense is complemented by the presence of aposematic or warning colorations. Opisthobranch species contain a vast source of different chemicals, many of which have been isolated for the first time from these animals. Some of these chemicals are being surveyed by the National Cancer Institute and private corporations for potential pharmaceutical use, and several are in clinical trials. For example, Dolastatin 10, an agent

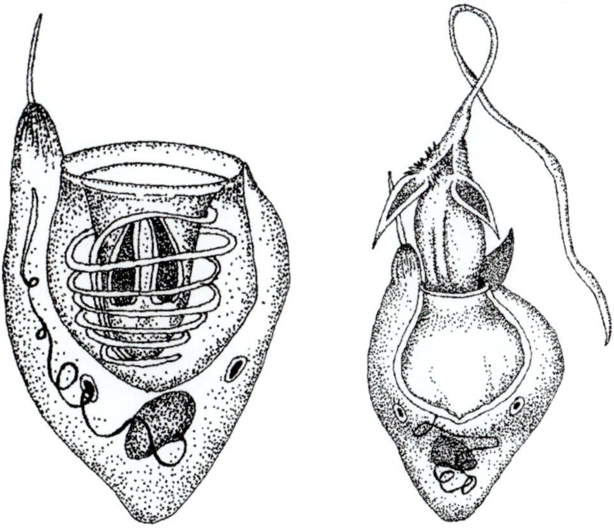

Stinging cells (nematocysts) ready to fire, left, and fired, right (drawings D. Behrens)

that looks promising against some types of cancer was originally extracted from the Pacific species *Dolabella auricularia*. Opisthobranchs also provide many of the basic models for neurobiological studies because of the large size of their individual nerve cells.

The other common defense system used by sea slugs is a platoon of capsules containing stinging cells located in the cerata of aeolid nudibranchs. Armed and ready to fire in the defense of the slug, the stinging cells, or nematocysts were ingested while feeding on the tissue of their prey species in the phylum Cnidaria, which include hydroids, anemones and corals. Once sequestered in the cnidoblast at the tip of each cerata, immature nematocysts that have passed through the digestive system, continue to develop into functional, friable, mature stinging cells. Both of these defense strategies as well as others are discussed in Behrens (2005).

Because of all these defensive systems, most species of opisthobranchs lack natural predators. However, sea slugs are often parasitized by other organisms, particularly parasitic copepods. These are internal parasites and their presence is only given away when the egg masses of the copepod protrude through the skin of the slug.

Copepod parasites on two species of Caribbean nudibranchs (photos J. Hamann)

Aside from our desire to find and photograph them, sea slugs are intrinsically valuable to humans. We refer you to Behrens (2005) for the full story.

This book

In the following pages you will find descriptions and photographs of most species of sea slugs described from the Caribbean Sea and adjacent areas. Some of these species are in need of study using modern techniques and consequently some of the identifications are provisional, until more information on the anatomy of the species becomes available. Other species are undescribed and therefore they do not have a full scientific name. Finally, some species were so poorly described by the original authors that they are impossible to recognize, and they are not included in this book. We consider this book to be work in progress and we expect to have to make changes and revisions in the near future. Some groups, particularly shell-bearing species are poorly represented and those chapters will need to be expanded. The end of the book contains references to the articles and books used for identifications and a glossary of scientific terms.

SPECIES DESCRIPTIONS

TABLE OF CONTENTS

"Lower Heterobranchia"
Family Aplustridae 9
Family Acteonidae 12

Cephalaspidea
Family Cylichnidae 16
Family Retusidae 20
Family Haminoeidae 24
Family Bullidae 30
Family Aglajidae 32
Family Gastropteridae 40
Family Runcinidae 44

Sacoglossa
Family Cylindrobullidae 46
Family Juliidae 48
Family Volvatellidae 50
Family Oxynoidae 52
Family Boselliidae 56
Family Placobranchidae 58
Family Limapontiidae 78
Family Hermaeidae 86
Family Caliphyllidae 88

Aplysiomorpha
Family Aplysiidae 94

Umbraculoidea
Families Umbraculidae & Tylodinidae 106

Pleurobranchomorpha
Family Pleurobranchidae 108

Nudibranchia
Family Hexabranchidae 116
Family Aegiretidae 118
Families Corambidae & Onchidorididae 122
Family Goniodorididae 124
Family Polyceridae 132
Family Chromodorididae 144
Family Dorididae 168
Family Discodorididae 174
Family Dendrodorididae 196
Family Phyllidiidae 206
Family Arminidae 208
Family Proctonotidae 210
Family Dotidae 212
Family Tritoniidae 222
Family Bornellidae 226
Family Lomanotidae 228
Family Scyllaeidae 230
Families Hancockidae & Tethydae 234
Family Flabellinidae 236
Family Eubranchidae 244
Family Facelinidae 248
Family Glaucidae 262
Family Tergipedidae 264
Family Aeolidiidae 270

Family APLUSTRIDAE Gray, 1847

Diagnosis: Body oval, partially covered by an external shell. Shell with thin walls, globose or ovoid, with a visible spire. Animal cannot completely retract into the shell. Cephalic shield large with a pair of broad posterior lobes. Anterior edge of the body folded into two pairs of enrolled lobes with the Hancock's organs situated below. Shell posteriorly surrounded by a projection of the mantle that folds up around and extends beyond the apex of the shell. Parapodia and operculum absent. Jaws composed of denticulate chitinous rods. Radula with or without a small central tooth with an anterior cusp and several large, triangular lateral teeth. Large mantle cavity opens to the right and contains a large gill.

Biology: Species of *Hydatina* and *Micromelo* feed on polychaete worms. Members of these genera occur in intertidal and shallow subtidal waters, where they crawl about in the open on reefs, sand, and seagrass flats.

Hydatina physis
(Linnaeus, 1758)

"Brown-line paperbubble"

Size: Up to 50 mm long

photo A. Valdés - shell, Florida

Micromelo undatus
(Bruguière, 1792)

"Miniature melo"

Size: Up to 30 mm long

photos A. Valdés & J. Hamann - left: specimen, Manzanillo, Costa Rica; right: egg mass, St. Lucia

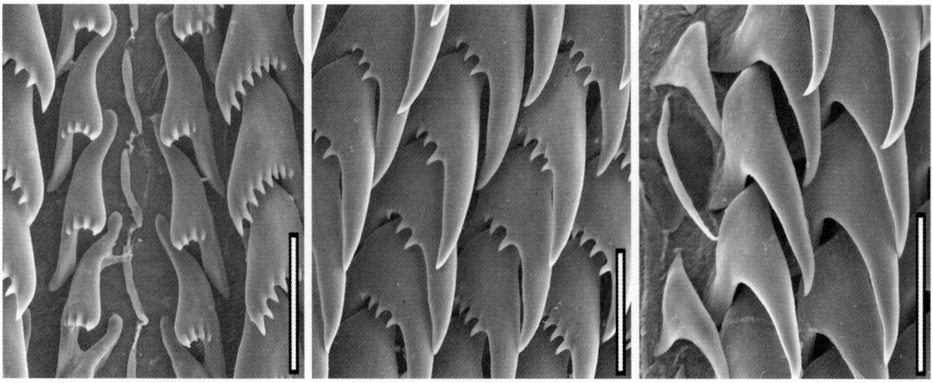

photos A. Valdés - SEM of *Micromelo undatus*; left: central radular teeth, scale bar = 50 µm; center: mid radular teeth, scale bar = 50 µm; right: outer radular teeth, scale bar = 50 µm

---

Bulla physis Linnaeus 1758: 727.
[several other synonyms from this and other geographic areas not listed]

Description: Background color reddish brown. Shield lobes and mantle projections with a bluish white marginal band. Shell oval, cream with thin black lines, covering only a small portion of the body.

Geographic Range: Circumtropical; Western Atlantic: Florida, Mexico, Colombia, Venezuela, Bermuda, Cuba, Jamaica, Virgin Islands, French Guyana.

Natural History: Inhabits sandy areas; feeds only on cirratulinid polychaete worms.

Bulla undata Bruguière 1792: 380.
Bulla nitidula Dillwyn 1817: 483.
Bullina elegans Menke 1830: 13, 125.

Description: Background color translucent gray with large, oval, opaque white spots all over the body. Body, cephalic shield lobes, and mantle projections with a greenish submarginal band, gradually becoming whitish, orange and finally bright red towards the edge of the body. Shell oval, shiny white with a network of thin dark red lines, covering only a small portion of the body.

Geographic Range: Circumtropical; Western Atlantic: Florida, Honduras, Costa Rica, Panama, Colombia, Curaçao, Bonaire, Venezuela, Bermuda, Cuba, Jamaica, Haiti, Puerto Rico, Virgin Islands, Martinique, St. Lucia, Barbados, St. Vincent & the Grenadines, Brazil.

Natural History: Inhabits sandy areas; feeds only cirratulinid polychaete worms.

Family ACTEONIDAE d'Orbigny, 1843

Diagnosis: Body short, wide, covered by a solid, external shell, fully able to contain the retracted animal inside. Shell spire usually conical, with up to eight whorls and the protoconch (larval shell) visible on top; columella with one or more strong folds; sculpture composed of punctate grooves; periostracum thin. Aperture normally with a chitinous operculum. Cephalic shield bilobed anteriorly; posteriorly with two large lobes that can cover the front part of the shell. Foot usually broad and short, with angular or tentacular anterior corners, and a posterior end normally shorter than the shell. Parapodia absent. Jaws composed of denticulate chitinous rods. Radula extremely variable, with or without a small median tooth, and a variable number of lateral teeth with different shapes within the same row, or all similar in shape.

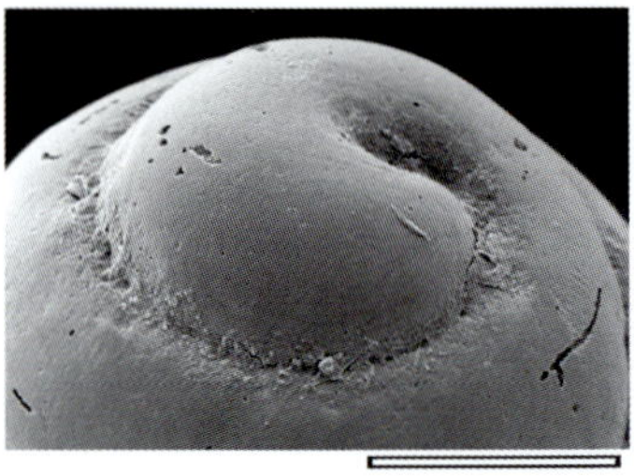

photos A. Valdés - SEM of *Japonacteon punctostriatus*; left: adult shell; right: larval shell, scale bar = 200 μm

Biology: Acteonids probably feed on polychaete worms. Members of this family are found in shallow and deep waters reaching depths exceeding 3,000 m. There is a rich diversity of deep-sea species which remain poorly known.

Mysouffa cumingii
(A. Adams, 1855)

Size: Up to 20 mm long

photos J. Hamann - Union Island, St. Vincent & the Grenadines; left: specimen; right: egg mass

Remarks: Several deep-water species of Acteonidae have been described from the tropical western Atlantic based on dredged empty shells. These species are not included in this book; they are: *Acteon danaida* Dall, 1881 (Dry Tortugas), *Acteon delicatus* Dall, 1889 (Barbados and Cuba), *Acteon incisus* Dall, 1881 (Yucatan Strait), *Acteon melampoides* Dall, 1881 (Cuba), *Acteon perforatus* Dall, 1881 (Cuba), *Acteon turritus* Watson, 1883 (Puerto Rico), and *Crenilabium exile* (Jeffreys, 1870) (Iceland to Mexico). Generic placement for members of this family requires examination of the radula, therefore the names of these species may require changes once complete specimens become available for study.

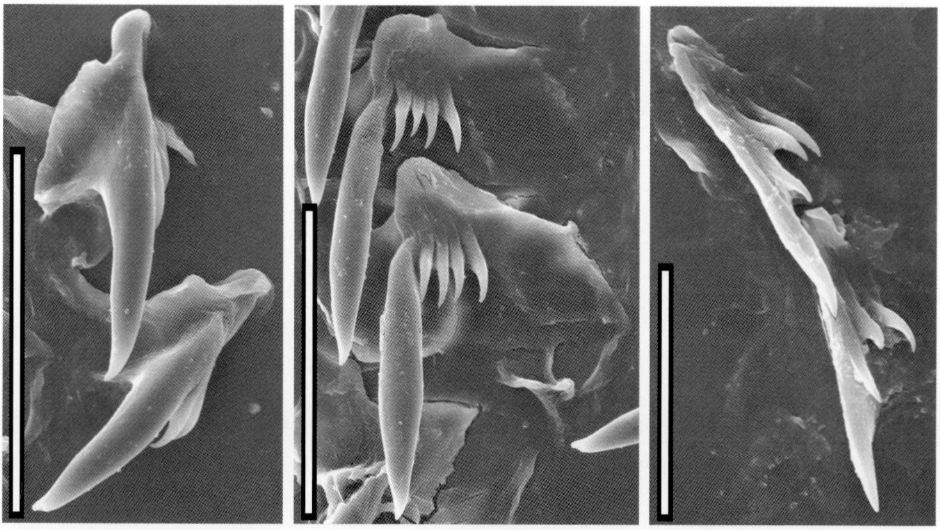

photos A. Valdés - SEM of *Japonacteon punctostriatus*; left: inner radular teeth, scale bar = 50 µm; center: mid radular teeth, scale bar = 50 µm; right: outer radular teeth, scale bar = 25 µm

Actaeon cumingii A. Adams 1855: 59.

Description: Background color uniformly white. Cephalic shield bilobed anteriorly, with two posterior, large lobes covering the front part of the shell. Shell large, solid, with a conspicuous sculpture composed of numerous punctate grooves covering the entire surface. Spire short, conical.

Geographic Range: Florida, Guadeloupe, St. Vincent & the Grenadines, Brazil.

photo © B. Frank - Florida

Acteon candens
Rehder, 1939

Size: Up to 10 mm long

photos A. Valdés - shells, Venezuela

Japonacteon punctostriatus
(C. B. Adams, 1840)

"Pitted baby-bubble"

Size: Shell up to 7.4 mm long

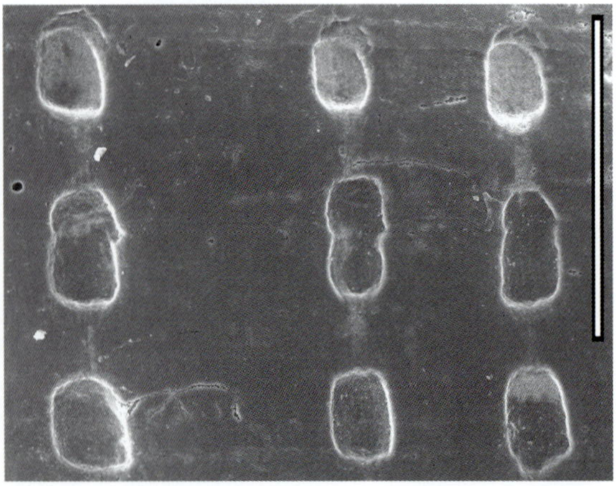

photo A. Valdés - shell sculpture, scale bar = 100 µm

Acteon candens Rehder 1939: 21, pl. 6, fig. 7.

Description: Background color uniformly white. Cephalic shield bilobed anteriorly, with two posterior, small lobes. Shell large, solid, with a faint sculpture composed of punctate grooves covering only the anterior end of the spire. Spire large, conical

Geographic Range: North Carolina, Florida, Louisiana, Texas, Costa Rica, Cuba, Brazil.

Tornatella punctostriata C. B. Adams 1840: 323.
Tornatella punctata d'Orbigny 1841: 230-231, pl. 17, figs. 10, 10', 11-12.
Actaeon cubensis Gabb 1873: 245.
Actaeon riomaensis Maury 1917: 175-177, pl. 3, fig. 1.

Description: Living animal unknown. Shell solid, elongate, with convex sides. Spire variable, conical, with 2-3 short to long whorls. Suture slightly channeled. Protoconch globose, about 1.5 whorl, 300 μm in diameter. Umbilicus absent. Aperture narrowing abruptly about its half-length. Anterior end of the aperture expanded into a small lip that is more conspicuous near the middle of the shell. Columellar margin thickened, slightly oblique, with a small, simple fold. Sculpture composed of a number of punctured spiral grooves. Punctuations oval, fused to each other within each groove or separate, absent from the posterior end of the spire. Grooves separated by gaps several times as wide as the grooves. Color whitish.

Geographic Range: Massachusetts, New York, New Jersey, Maryland, Virginia, North Carolina, South Carolina, Florida, Texas, Mexico, Costa Rica, Panama, Colombia, Venezuela, Bermuda, Cuba, Jamaica, Dominican Republic, Puerto Rico, Virgin Islands, Guadeloupe, Martinique.

Geographic Range: The radular morphology of this species resembles that of members of *Japonacteon*, therefore it is here transferred to this genus.

Family CYLICHNIDAE H. Adams & A. Adams, 1854

Diagnosis: Body oval to elongate with a strongly calcified, large external shell, able to contain the retracted animal. Shell oval pyriform or cylindrical, with or without a spire. Sculpture absent or faint in most cases, some species with punctuate grooves. Cephalic shield short with a pair of posterior lobes; foot short lacking parapodia. Digestive system normally with no jaws, radula usually with a median tooth and from one to several curved, generally denticulate, lateral teeth. Gizzard with three plates.

Biology: Species of Cylichnidae probably feed on a variety of marine organisms such as foraminiferans, annelids, crustaceans and other mollusks. They are found from intertidal mudflats to the deep-sea down to 5,000 m depth.

Remarks: This group is in need of work in the Caribbean and the identifications here provided are provisional. The species illustrated are an example of the diversity of the group, and several others are missing.

Cylichnella bidentata
(d'Orbigny, 1841)

Size: Up to 4 mm long

photos © B. Frank & © Y. Camacho - left: specimen, Florida (BF); right: shell, Costa Rica (YC)

Acteocina canaliculata
(Say, 1826)

Size: Up to 6 mm long

photo A. Valdés - Sanibel Island, Florida

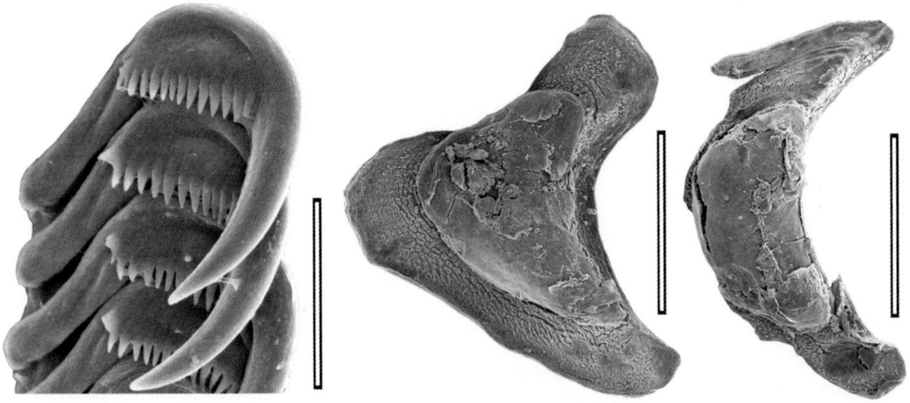

photos A. Valdés - SEM of *Acteocina candei*; left: radular teeth, scale bar = 20 µm; right: gizzard plates, scale bar = 300 µm

Bulla bidentata d'Orbigny 1841: 125-126, pl. 4, figs. 13-16.
?*Bulla biplicata* Lea 1845: 204.
Cylichnella mareana Weisbord 1962: 458-459, pl. 47, figs. 1-2.

Description: Background color uniformly dirty yellow. Shell translucent whitish with the viscera visible through it. Shell oval to elongate, with convex sides. Spire and umbilicus absent. Aperture as long as the shell. Columellar margin thickened, slightly oblique, with a small, simple fold. Sculpture composed of several spiral grooves near the anterior and posterior ends.

Geographic Range: New Jersey, North Carolina, Florida, Texas, Mexico, Panama, Colombia, Venezuela, Cuba, Jamaica, Puerto Rico, Virgin Islands, Guadeloupe, Martinique, Surinam, Brazil.

Volvaria canaliculata Say 1826: 211.
[other synonyms not included]

Description: Background color translucent white with some opaque white spots on the head. Head with two large posterior lobes, parapodia absent. Shell white, solid, oval to elongate. Spire short, conical, with 2-3 whorls. Suture channeled. Umbilicus absent. Columellar margin thickened, slightly oblique, with a small, simple fold.

Geographic Range: Nova Scotia, New Brunswick, Maine, Massachusetts, Rhode Island, Connecticut, New York, New Jersey, Maryland, Virginia, North Carolina, South Carolina, Georgia, Florida, Louisiana, Texas, Mexico.

photo © L. Harris - Bocas del Toro, Panama

Acteocina candei
(d'Orbigny, 1842)

Size: Up to 4.5 mm long

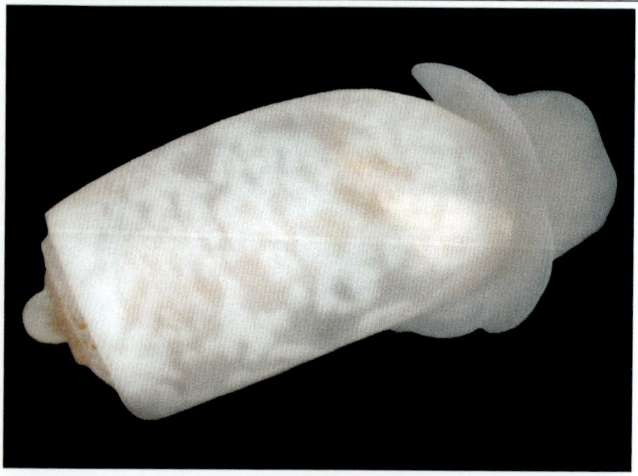

photo © C. Redfern - Abaco, Bahamas

Acteocina lepta
Woodring, 1928

"Woodring's barrel-bubble"

Size: Up to 5.3 mm long

photo © B. Frank - Florida

Acteocina atrata
P.S. Mikkelsen & P.M. Mikkelsen, 1984

"Blackback barrel-bubble"

Size: Up to 5.4 mm long

Bulla candei d'Orbigny 1841: 128-129, pl. 4 bis, figs. 1-4.
[other synonyms not included]

Description: Background color translucent white. Shell translucent white with the viscera visible as an irregular pattern of white pigment on an orange background. Shell solid, elongate. Spire long, conical, with channeled 2-3 whorls. Umbilicus absent. Columellar margin thickened, slightly oblique, with a small, simple fold.

Geographic Range: North Carolina, Florida, Texas, Mexico, Costa Rica, Panama, Colombia, Venezuela, Bermuda, Cuba, Jamaica, Puerto Rico, Virgin Islands, Guadeloupe, Martinique, Guyana, Brazil, Argentina.

Acteocina lepta Woodring 1928: 121, pl. 2, fig. 5.

Description: Background color translucent white. Head with two large posterior lobes, parapodia absent. Shell translucent white with the viscera visible as an irregular pattern of white pigment on an a reddish background. Shell solid, oval to elongate. Spire very short, almost flat, with 2-3 whorls. Umbilicus absent. Columellar margin thickened, slightly oblique, with a small, simple fold.

Geographic Range: North Carolina, South Carolina, Georgia, Florida, Louisiana, Texas, Mexico, Bermuda, Bahamas, Cuba, Cayman Islands, Jamaica, Dominican Republic, Puerto Rico, Barbados, Brazil.

Acteocina atrata P.S. Mikkelsen & P.M. Mikkelsen 1984: 179-187, fig. 7A-B, 8A-G, 9A-C, 10A-F.

Description: Background color translucent gray with some reddish pigment on the head. Head with two large posterior lobes, parapodia absent. Shell white, solid, oval. Spire short, conical, with 2-3 whorls. Suture channeled. Umbilicus absent. Columellar margin thickened, slightly oblique, with a small, simple fold.

Geographic Range: Florida.

Family RETUSIDAE Thiele, 1925

Diagnosis: Body elongate, lacking parapodia. Cephalic shield with a shallow notch in front and two lobes posterio-laterally, which normally do not cover any part of the shell. Eyes visible at the base of the anterior end of the shell. Shell covering the body almost completely; the shell holds the retracted animal inside. Shell cylindrical or pear-shaped; posterior end normally truncate or sunken, in some groups with a conical extension or a short spire; protoconch (larval shell) normally visible. Shell surface often with a sculpture composed of spiral grooves, growth lines, or both, in most cases spiral lines more conspicuous near the anterior and posterior ends of the shell. Operculum present in some species. Digestive system with no jaws or radula. Gizzard with three oval plates, one of them usually larger than the others, except for *Volvulella*, which lacks gizzard plates completely.

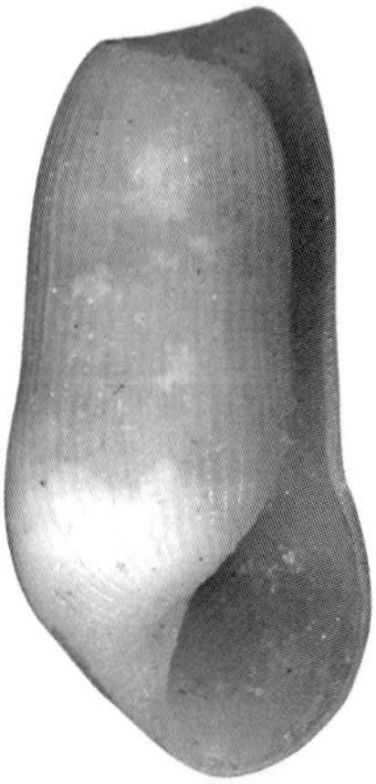

Retusa sulcata
(d'Orbigny, 1841)

"Sulcate barrel-bubble"

Size: Shell up to 3 mm long

photo A. Valdés - shell, Jamaica

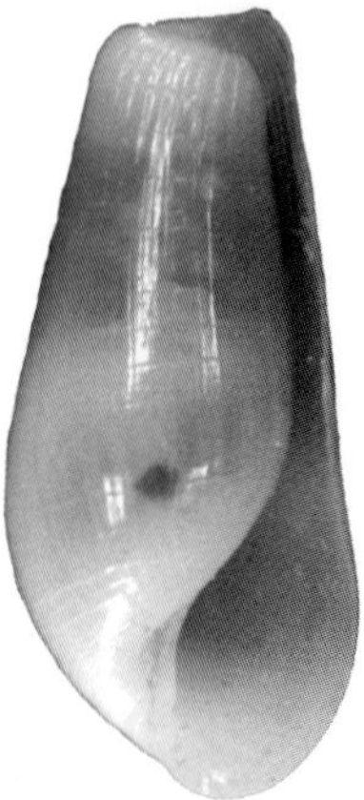

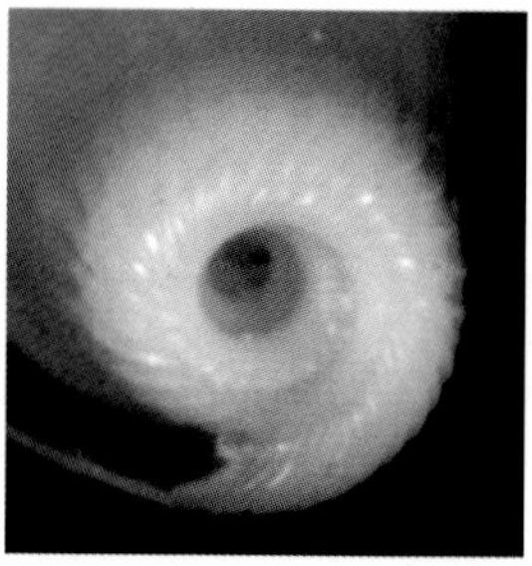

Retusa caelata
(Bush, 1885)

Size: Shell up to 4.2 mm long

photo A. Valdés - shell, Florida

Biology: Members of this family are normally small or very small infaunal species (species that live buried in sediment). They are found from shallow intertidal sand and mud flats to abyssal depths greater than 5,000 m. Retusids appear to feed on diatoms, foraminiferans and small molluscs.

Remarks: Deep water species of Retusidae are often confused with juvenile Bullidae. Anatomical examination of some deep-sea western Atlantic *Bulla* species may eventually show that they are members of *Roxania* or other retusid genera; the deep water species are not included in this book. Despite their local abundance and because of their small size, these animals are difficult to find alive and photograph, therefore for some of the species treated in this chapter only the shells are illustrated and described.

Bulla sulcata d'Orbigny 1841: 129-130, pl. 4 bis, figs. 9, 21.
?*Retusa omphalis* Mörch 1874: 172

Description: Living animal unknown. Shell fragile, elongate, wider anteriorly and narrower posteriorly, with nearly parallel or slightly concave sides. Apex umbilicated, with the aperture lip rising from the right side. Aperture lip forms a rounded wing connected to the columellar margin. Anterior end of the shell rounded. Aperture as long as the shell. Umbilicus absent. Sculpture with a number of conspicuous growth lines crossed by faint spiral lines. Color uniformly whitish.

Geographic Range: North Carolina, Florida, Louisiana, Texas, Costa Rica, Panama, Colombia, Venezuela, Cuba, Jamaica, Virgin Islands, Guadeloupe.

Cylichna caelata Bush 1885: 468, pl. 45, fig. 15.
Retusa iheringi von Ihering 1915: 138.

Description: Living animal unknown. Shell fragile, elongate, wider anteriorly and narrower posteriorly. Apex umbilicated, with the aperture lip rising from the right side. Aperture lip forms a rounded wing connected to the columellar margin. Anterior end of the shell rounded. Aperture as long as the shell. Umbilicus absent. Sculpture with a number of growth lines more conspicuous near the posterior end of the shell, crossed by well-marked spiral lines more densely arranged near the anterior and posterior ends of the shell. Color uniformly whitish.

Geographic Range: North Carolina, Florida, Louisiana, Texas, Mexico, Venezuela, Puerto Rico, Guyana, Surinam, Brazil, Uruguay, Argentina.

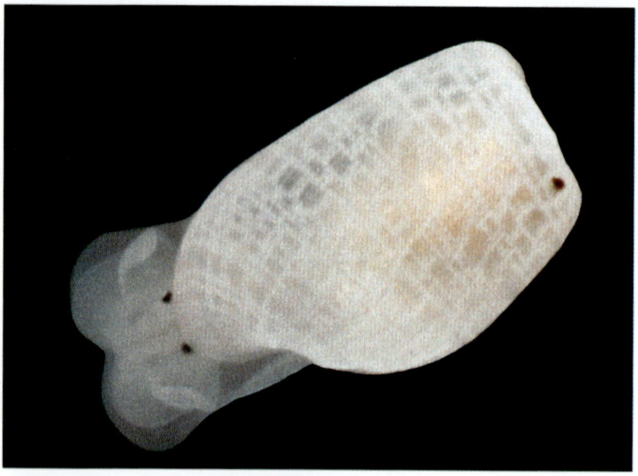

Retusa alayoi
(Espinosa & Ortea, 2004)

Size: Shell up to 2 mm long

photo © C. Redfern - Abaco, Bahamas

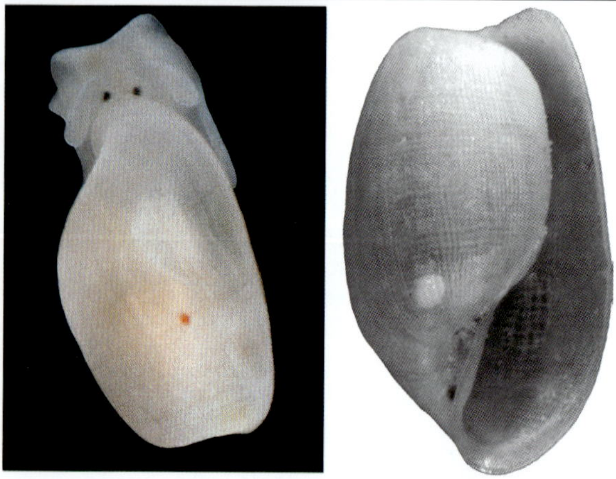

***Retusa* sp. 1**

Size: Shell up to 3.5 mm long

photos © C. Redfern & A. Valdés - left: specimen, Bahamas (CR); right: shell, Venezuela (AV)

Volvulella acuta
(d'Orbigny, 1841)

Size: Shell up to 4.6 mm long

photos A. Valdés - left: shell, Costa Rica; right: shell, Jamaica

Atys alayoi Espinosa & Ortea 2004: 213-214, fig. 4A-B; pl. 1, fig. E.

Description: Background color uniformly translucent white. Cephalic shield with two lateral lobes and a frontal notch; parapodia absent. Shell fragile, quadrangular, wider near the center, with convex sides. Apex umbilicated, with the aperture lip rising from the right side. Aperture lip forms a rounded wing connected to the columellar margin. Anterior end of the shell rounded. Aperture as long as the shell. Umbilicus very narrow. Sculpture with a number of large, rectangular punctuations. Color translucent whitish.

Geographic Range: Cuba, Bahamas.

Remarks: The external morphology of the living animal has features characteristic of the Retusidae; this species is here assigned to the genus *Retusa* until anatomical studies are available.

Description: Background color translucent white. Cephalic shield with two lateral lobes and a frontal notch; parapodia absent. Shell fragile, oval, wider near the center, with convex sides. Apex umbilicated, with the aperture lip rising from the right side. Aperture lip forms a rounded wing connected to the columellar margin. Anterior end of the shell rounded. Aperture as long as the shell. Umbilicus very narrow. Sculpture with a number of simple spiral lines crossed by conspicuous growth lines. Color uniformly whitish.

Geographic Range: Venezuela, Bahamas.

Remarks: This appears to be an undescribed species.

Bulla acuta d'Orbigny 1841: 126, pl. 4, figs. 17-20.
[other synonyms not included]

Description: Living animal unknown. Shell fragile, very variable in shape, from short and oval to narrow and elongate, with nearly parallel or convex sides. Anterior end of the shell rounded. Umbilicus absent. At its apical end, the aperture bends over the apex of the shell and the parietal wall forms a conical spine that completely covers the apex. Sculpture composed of a number of irregular spiral grooves crossed by numerous fine growth lines all over the shell surface. Spiral grooves situated near the anterior and posterior ends of the shell more conspicuous. Color uniformly shiny whitish.

Geographic Range: Costa Rica, Venezuela, Curaçao, Cayman Islands, Barbados, Jamaica.

Family HAMINOEIDAE Pilsbry, 1895

Diagnosis: Body oval to elongate, with short parapodia partially covering the shell. Cephalic shield with the anterior corners folded down ventrally. Shell ovoid to cylindrical, covering only a portion of the body, but it can hold the entire retracted animal. Shell thin, fragile (except for a couple of species of *Atys*) and translucent; usually smooth, with or without fine spiral lines, and with a very thin periostracum. Shell spire sunken; columella normally smooth without folds. Penis sometimes armed with chitinous spines. Jaws composed of chitinous rods, sometimes arranged in polygonal bundles. Radula with a central tooth and from a single to numerous hook shaped lateral teeth. Gizzard with three identical plates with transverse lines of denticles on their inner side.

Biology: Haminoeids are herbivorous, feeding on green algae. Many species of *Haminoea* crawl on mud and sand flats at low tide with very little or no water. Some species of *Atys* are nocturnal and emerge at night to feed.

Haminoea elegans
(Gray, 1825)

"Elegant glassy-bubble"

Size: Up to 35 mm long

photo J. Hamann - St. Lucia

Haminoea antillarum
(d'Orbigny, 1841)

"Antilles glassy-bubble"

Size: Up to 40 mm long

photos A. Valdés - Sanibel Island, Florida; left: adult specimen; right: egg mass

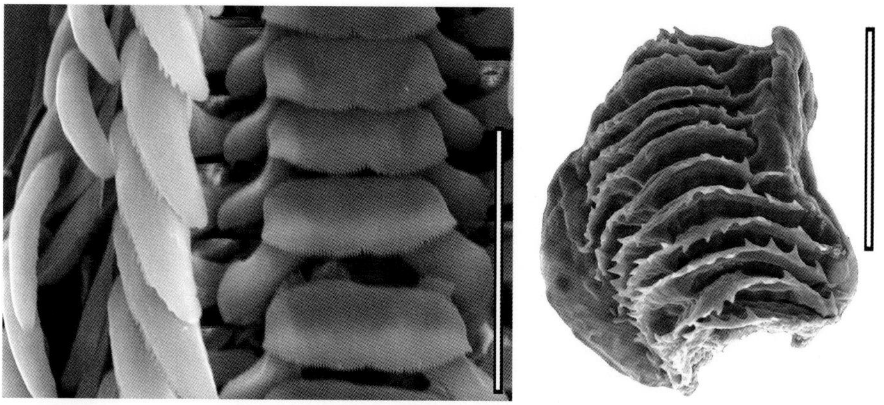

photos A. Valdés - SEM of *Atys caribaeus*; left: radular teeth, scale bar = 20 μm; right: gizzard plate, scale bar = 50 μm

Bulla elegans Gray 1825: 408.
Bulla diaphana Gould 1852: 222.
Bullae guildingii Swainson 1840: 360, fig. 46.
Haminoea taylorae Petuch 1987: 31-32, pl. 4, figs. 12-13.

Description: Background color translucent yellowish gray, with numerous black and opaque white spots. Parapodia large, covering the anterior part of the shell. Body elongate, with a long, deeply notched cephalic shield having reduced tentacles. Shell fragile, lightly calcified, with numerous and conspicuous spiral grooves crossed by growth lines.

Geographic Range: Florida, Texas, Mexico, Belize, Honduras, Costa Rica, Colombia, Curaçao, Bonaire, Venezuela, Bermuda, Cuba, Jamaica, Puerto Rico, Virgin Islands, Martinique, St. Lucia, St. Vincent & the Grenadines, Grenada, Trinidad & Tobago, Brazil.

Bulla antillarum d'Orbigny 1841: 124-125, pl. 4, figs. 9-12.
Bulla (Haminea) cerina Menke 1853: 142.
Haminea guadaloupensis Sowerby II 1868: pl. 3, fig. 14.

Description: Background color translucent gray with dense concentrations of black pigment, opaque white dots and distinctive orange spots all over the body; orange spots visible through the shell. Body wide, with a short and broad cephalic shield having well-defined lateral tentacles. Parapodia cover only the anterior end of the shell. Posterior end of the foot very broad and rounded. Shell translucent, smooth, with no visible sculpture, except for faint growth lines.

Geographic Range: Florida, Texas, Mexico, Honduras, Panama, Colombia, Venezuela, Bermuda, Cuba, Cayman Islands, Jamaica, Puerto Rico, Virgin Islands, Guadeloupe, Brazil.

Haminoea succinea
(Conrad, 1846)

"Amber glassy-bubble"

Size: Up to 20 mm long

photo A. Valdés - Sanibel Island, Florida

Atys caribaeus
(d'Orbigny, 1841)

"Caribbean glassy-bubble"

Size: Up to 20 mm long

photos J. Hamann - left: Nicholas, Belize; right: Carriacou, St. Vincent & the Grenadines

Atys macandrewii
Smith, 1872

Size: Up to 10 mm long

photo © C. Redfern - Abaco, Bahamas

Bulla succinea Conrad 1846: 26.
Haminea succinea var. *solidior* Vanatta 1901: 182-183.

Description: Background color translucent gray with numerous black spots and some opaque white dots. Body elongate, with a broad cephalic shield having two lateral, very short and wide tentacles. Parapodia very short, not covering any portion of the shell. Shell fragile, elongate, translucent white.

Geographic Range: Florida, Louisiana, Texas, Mexico, Colombia, Venezuela, Bermuda, Puerto Rico, St. Maarten/St. Martin, St. Barthelemy.

Bulla caribaea d'Orbigny 1841: 127, pl. 4, figs. 21-24.
Atys sharpi Vanatta 1901: 183, pl. 5, fig. 9.

Description: Background color translucent white with irregular opaque white spots and some black dots. Body very elongate, with the cephalic shield deeply notched posteriorly and two short parapodia covering the anterior end of the shell. Shell elongate, translucent, with conspicuous spiral grooves near the anterior and posterior ends.

Geographic Range: North Carolina, Florida, Mexico, Belize, Costa Rica, Panama, Colombia, Venezuela, Cuba, Jamaica, Puerto Rico, Virgin Islands, St. Maarten/St. Martin, Martinique, St. Lucia, Martinique, Guadeloupe, St. Vincent & the Grenadines, Grenada, Brazil.

Atys m'andrewii Smith 1872: 346.
Atys lineata Usticke 1959: 85, pl. 4, fig. 15.

Description: Background color translucent white with some opaque white spots. Body very elongate, with the cephalic shield deeply notched posteriorly and two short parapodia covering the anterior end of the shell. Large orange spots visible through the shell. Shell elongate, with distinctive dark spiral lines all over the surface.

Geographic Range: Amphiatlantic; Western Atlantic: Louisiana, Costa Rica, Curaçao, Bahamas, Virgin Islands, Brazil.

photo J. Hamann - Crocus Bay, Anguilla

Atys riiseanus
Mörch 1875

Size: Up to 20 mm long

photo J. Hamann - Flamingo Bay, Virgin Islands

Atys guildingii
Sowerby II, 1869

Size: Up to 30 mm long

photo A. Valdés - Bocas del Toro, Panama

Atys sandersoni
Dall, 1881

Size: Up to 15 mm long

Atys riiseana Mörch 1875: 173.

Description: Background color translucent white with densely arranged black dots on the head and parapodia. Irregular brown patches of the digestive gland visible through the shell. Body short and wide, with the parapodia covering the anterior end of the shell. Shell oval, short, with conspicuous spiral grooves near the anterior and posterior ends of the shell.

Geographic Range: Florida, Mexico, Costa Rica, Panama, Colombia, Curaçao, Bahamas, Cuba, Dominican Republic, Puerto Rico, Virgin Islands, Anguilla, St. Maarten/St. Martin, Guadeloupe, Trinidad & Tobago, Brazil.

Atys guildingii Sowerby II 1869: pl. 5, fig. 26A-B.

Description: Background color translucent white with densely arranged brown dots on the head and parapodia. Brown areas of the digestive gland visible through the shell. Body short and wide, with the parapodia covering the anterior end of the shell. Shell oval, short, with faint spiral grooves near the anterior and posterior ends of the shell.

Geographic Range: Mexico, Costa Rica, Panama, Colombia, Venezuela, Puerto Rico, St. Vincent & the Grenadines, Brazil.

Atys (?) sandersoni Dall 1881: 99.

Description: Background color translucent white with dense black pigment on the head and parapodia. Body very elongate, with the cephalic shield notched posteriorly and two short parapodia covering the anterior end of the shell. Shell elongate, translucent, with no conspicuous sculpture.

Geographic Range: North Carolina, Florida, Mexico, Costa Rica, Panama, Cuba, Puerto Rico, Brazil.

Family BULLIDAE Gray, 1827

Diagnosis: Body elongate, completely covered by a strong, inflated shell with a deeply apical umbilicus or involute spire. Animal can retract completely into the shell. Shell surface very smooth, spotted, aperture at least as long as the shell; columella with no ridges or folds, but with a thin columellar callus. Cephalic shield with a pair of inrolled siphon-like tentacles anteriorly and a pair of broadly triangular lobes covering the anterior part of the shell posteriorly. Parapodia small. Jaws composed of closely packed chitinous rods. Radula with a broad, denticulate central tooth and an inner symmetrical and an outer asymmetrical denticulate lateral tooth. Gizzard plates with chitinous spines.

Biology: Members of this family are nocturnal, and feed on algae during the hours of darkness, burrowing or hiding away during the day. They are shallow water animals, often in estuarine or lagoonal environments, and prefer areas of soft sandy mud.

Bulla striata
Bruguière, 1792

"Striate bubble"

Size: Up to 65 mm long

photos J. Hamann - left: Simpson Lagoon, St. Maarten/St. Martin; right: Piton Bay, St. Lucia

Bulla occidentalis
A. Adams, 1850

Size: Up to 50 mm long

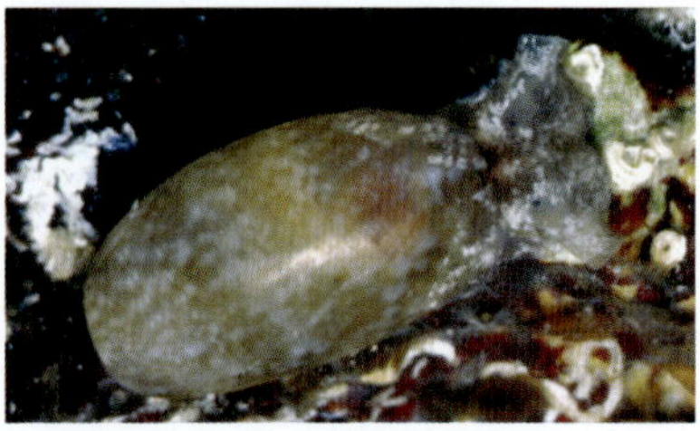

photos A. Valdés - Sanibel Island, Florida; left: adult specimen; right: juvenile specimen

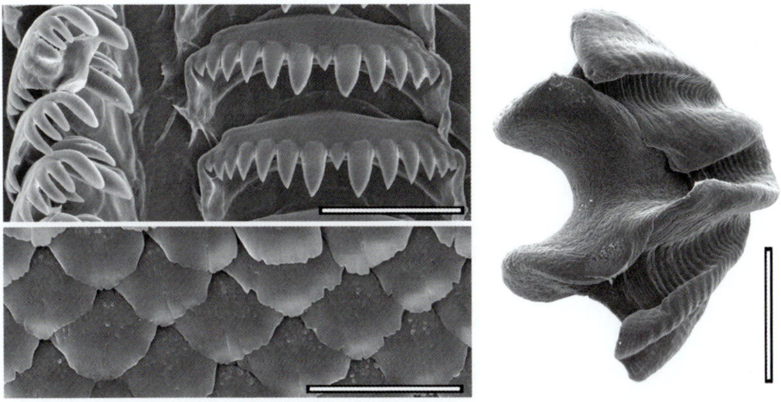

photos A. Valdés - SEM of *Bulla striata*; upper left: radular teeth, scale bar = 100 µm; right: gizzard plate, scale bar = 300 µm; lower right: jaw elements, scale bar = 20 µm

Bulla striata Bruguière 1792: 372-373.
[other synonyms from this and other geographic areas not included]

photo J. Hamann - Guadeloupe

Description: Background color translucent gray with opaque white spots particularly behind the eyes. Parapodia and tentacles long. Shell elongate in profile, narrower in the middle, slightly concave on the left side, with strong spiral lines near both ends. Shell brownish, normally mottled with black, brown and white.

Geographic Range: Amphiatlantic; Western Atlantic: North Carolina, Florida, Louisiana, Texas, Mexico, Belize, Costa Rica, Colombia, Venezuela, Bermuda, Curaçao, Cuba, Jamaica, Puerto Rico, Virgin Islands, Bahamas, Guadeloupe, Martinique, Dominica, St. Maarten/St. Martin, St. Vincent & the Grenadines, Grenada, Barbuda, Antigua, St. Lucia, Guadeloupe, Barbados, Brazil.

Bulla occidentalis A. Adams 1850: 577, pl. 123, fig. 72-73.

Description: Background color translucent gray with numerous opaque white spots on the head and parapodia as well as some brown pigment. Parapodia and tentacles short. Shell oval in profile, wider in the middle, with both ends rounded and faint spiral lines near both ends. Shell brownish, mottled with dark brown and white.

Geographic Range: Florida, Texas, Mexico, Honduras, Panama, Colombia, Venezuela, Cuba, Virgin Islands, Barbuda, Antigua, St. Vincent & the Grenadines.

Natural History: Inhabits areas with sand and sea grass; it buries during day and becomes active at night.

Family AGLAJIDAE Pilsbry, 1895

Diagnosis: Body elongate, narrow, with a large cephalic shield, a posterior visceral hump and two parapodia. Head with a series of sensory bristles next to the mouth. Cephalic shield with a posterior lobe sometimes raised vertically or overlapping the anterior end of the visceral hump. Visceral hump posteriorly projecting into one or more caudal lobes, more or less elongate, resembling tails. Shell normally reduced to an internal plate, translucent, membranous or well calcified. Mantle cavity opened laterally, with a single gill. Digestive system without jaws and gizzard plates and normally without radula.

Biology: Aglajids are active predators on opisthobranchs, polychaete and nemertean worms, or flat worms. Some species display complex mating behaviors; others track and follow members of the same species and communicate danger using pheromones.

photo A. Valdés - SEM of the shell of *Philinopsis pusa*, scale bar = 20 µm

---

photo J. Hamann - Isla Mujeres, Mexico

Philinopsis pusa
(Ev. Marcus & Er. Marcus, 1966)

Size: Up to 45 mm long

photos J. Hamann - Store Bay, Tobago; left: dorsal view; right: ventral view

photos A. DuPont - Behavior and morphology of *Aglaja hirundinina*; left: specimens mating; center: specimens trailing or tailgating; right: detail of the sensory bristles on the head.

Aglaja pusa Ev. Marcus & Er. Marcus 1966: 170.
?*Aglaja quinza* Ev. Marcus 1979: 132-133, figs. 1-4.

Description: Background color variable from pale translucent gray to dark brown. Dorsum covered with dark or white spots and yellow patches on the anterior end posterior ends of some specimens. Mantle margin with a blue edge in dark animals. Posterior end of the cephalic shield with an erect triangular lobe. Posterior end of the body with two equal, rounded lobes.

Geographic Range: Florida, Mexico, Tobago.

Natural History: Feeds on other headshield slugs. Inhabits shallow sand and buries in daytime.

Remarks: Very similar to the Mediterranean species *Aglaja depicta* Renier, 1807, it is distinguishable by the morphology of the penis (T. Gosliner, personal communication).

photos J. Hamann - left: Guadeloupe; right: Isla Mujeres, Mexico

Philinopsis petra
(Ev. Marcus, 1976)

Size: Up to 22 mm long

photo J. Hamann - Trou Madame, Martinique

photos J. Hamann - left: Cayman Islands; right: Martinique

Aglaja felis
Er. Marcus & Ev. Marcus, 1970

Size: Up to 12 mm long

photos J. Hamann - left: Long Point, Bahamas; right: Cayman Islands

Chelidonura petra Ev. Marcus, 1976b: 13-16, figs. 3-8.
Philinopsis aeci Ortea & Espinosa 2001: 41, pl. 2, fig. C, pl. 3, fig. A.

Description: Background color variable, from light yellow to gray, green or dark brown. Dorsum with an irregular pattern of yellow and white pigment. Parapodia wide, with a swelling at mid-length. Posterior end of the body with two short triangular lobes. Head with two lateral projections.

Geographic Range: Belize, Honduras, Costa Rica, Martinique, Cayman Islands, Bahamas, Brazil.

Natural History: Inhabits shallow areas and buries in sand.

Remarks: Anatomical examination of the material illustrated here confirmed that *Philinopsis aeci* is a synonym of *Chelidonura petra*, originally described based on a preserved specimen.

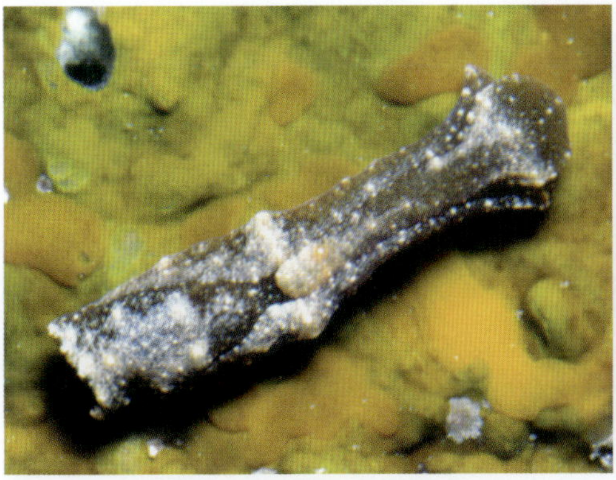

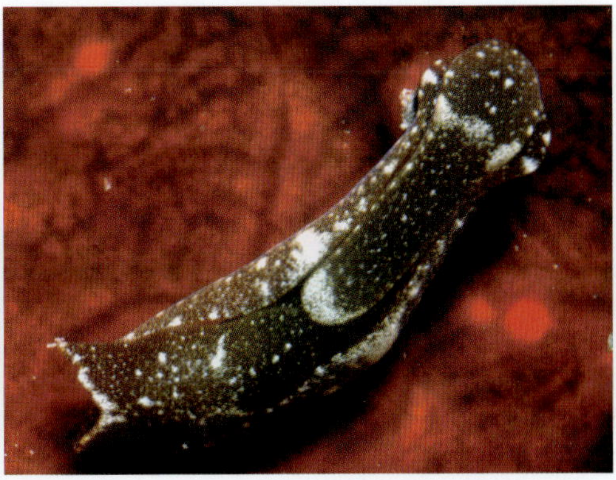

photos J. Hamann - Roatán, Honduras

Aglaja felis Er. Marcus & Ev. Marcus 1970: 9-10, figs. 1-3.

Description: Background color velvet black. Some specimens with a pattern of white lines and patches on the dorsum. Parapodia very short. Cephalic shield much longer than the visceral hump. Posterior end of the visceral hump with two uneven posterior, rounded lobes.

Geographic Range: Belize, Honduras, Colombia, Cayman Islands, Bahamas, Jamaica, Puerto Rico.

Natural History: Common on shallow sandy areas; it buries in sand.

photo J. Hamann - Drowned Key, Belize

Aglaja sp.

Size: Up to 16 mm long

photo J. Hamann - Guanaja, Honduras

Chelidonura berolina

Er. Marcus & Ev. Marcus, 1970

Size: Up to 12 mm long

photos J. Hamann - left: Harrington Sound, Bermuda; right: Isla Mujeres, Mexico

Description: Background color translucent gray with numerous opaque white spots all over the dorsum. Cephalic shield slightly longer than the visceral hump. Anterior end of the body rounded; posterior end with two extensions, left one long, narrow and pointed, right one rounded.

Geographic Range: Belize.

Natural History: This appears to be an undescribed species.

?*Aglaja hummelincki* Er. Marcus & Ev. Marcus 1970: 10-11, figs. 4-6.
Chelidonura berolina Er. Marcus & Ev. Marcus 1970: 17-18, figs. 18-19.
Chelidonura juancarlosi Ortea & Espinosa 1998: 4-5, figs. 1B-C.
Chelidonura mariagordae Ortea, Espinosa & Moro 2004: 218-219, figs. 1A-B, pl. 1, figs. A-D.

Description: Background color black with a submarginal yellow band on the parapodial edge and anterior and posterior ends of the body. In some specimens the band is interrupted and in others may be absent. Anterior end of the body translucent with the eyes visible. Dorsum covered by white and yellow patches in some specimens, with the posterior lobe of the cephalic shield always either white or yellow. Cephalic shield slightly longer than the visceral hump. Posterior end of the body with two elongate lobes, the left one being much longer.

Geographic Range: Amphiatlantic; Western Atlantic: Mexico, Belize, Honduras, Colombia, Cayman Islands, Cuba, Jamaica, Martinique, Puerto Rico, Bermuda.

Natural History: Common on shallow sandy areas; it buries in sand.

Remarks: Very similar to the eastern Atlantic species *Chelidonura africana* Pruvot-Fol 1953, which could be a synonym.

photos J. Hamann & T. Gosliner - left: Martinique (JH); right: Cayman Islands (TG)

Chelidonura hirundinina
(Quoy & Gaimard, 1833)

"Leech aglaja"

Size: Up to 20 mm long

photos J. Hamann - left: Guadeloupe, adult specimen; right: Cayman Islands, juvenile

Chelidonura cubana
Ortea & Martínez, 1997

Size: Up to 9 mm long

photos J. Hamann - Cayman Islands; left: adult specimen; right: juvenile specimen

Navanax aenigmaticus
(Bergh, 1893)

"Mysterious aglaja"

Size: Up to 50 mm long

photos J. Hamann - left: Ranguana Island, Belize; right: Roatán, Honduras

photo J. Hamann - San Andrés, Colombia

Bulla hirundinina Quoy & Gaimard 1832-33[1833]: 367-369, pl. 26, figs. 20-25. [several other synonyms not included].

Description: Background color variable from red or orange to dark brown or black. Dorsum with blue, black and orange stripes, and a conspicuous white patch on the posterior end of the cephalic shield. Posterior end of the body with a pair of tapering, elongate projections.

Geographic Range: Indo-Pacific, Western Atlantic: Florida, Belize, Colombia, Curaçao, Cayman Islands, Bahamas, Jamaica, Puerto Rico, Guadeloupe, St. Vincent & the Grenadines, Grenada.

Natural History: Feeds on flatworms, common on shallow sandy areas. It secretes a purple-brown liquid when disturbed.

Chelidonura cubana Ortea & Martínez 1997: 138-140, figs. 1, 2D.

Description: Background color dark brown to black with a pattern of blue spots on the edge of the parapodia and orange lines on the dorsum. Cephalic shield much longer than the visceral hump. Posterior end of the visceral hump with two uneven posterior, elongate and tapering lobes.

Geographic Range: Cayman Islands, Cuba.

Navarchus aenigmaticus Bergh 1894: 217-222, pl. 10, figs. 11-12, pl. 11, figs. 6-9, pl. 12, figs. 8-10. [several other synonyms not included].

Description: Background color from dirty yellow to dark brown. Dorsum covered with white and brown longitudinal lines and some whitish areas. Edge of the parapodia with a row of bright blue spots. Posterior end of the body with two lobes, the left one with a thin elongate projection.

Geographic Range: Eastern Pacific, Western Atlantic: Florida, Belize, Honduras, Costa Rica, Colombia, Venezuela, Bahamas, Curaçao, Jamaica, Virgin Islands, Guadeloupe, Martinique, Barbuda, St. Barthelemy, St. Vincent & the Grenadines, Grenada, Barbados, Brazil.

Natural History: Inhabits rocky areas and feeds on other opisthobranchs.

Remarks: *Posterobranchaea maculata* d'Orbigny, 1835 is probably the valid name for this species; we use here the most frequently used name until this question is clarified.

Family GASTROPTERIDAE Swainson, 1840

Diagnosis: Body oval to elongate, divided into a short, triangular cephalic shield and visceral hump, with a posterior medial siphon raised and recurved forward over the shield. Shell internal, reduced or absent. Visceral hump large, elongate with a right lateral ridge representing the upper free margin of the mantle cavity and in some cases additional terminal protuberances. Foot broad, longer than the visceral hump, developed laterally into two large, thin parapodia. Parapodia meet dorsally or overlap over the visceral hump. Pedal gland present in the posterior part of the sole and Hancock's organs in the lateral groove separating cephalic shield and foot. Penis opens on the right side behind the mouth. Mantle cavity reduced or absent. Jaws composed of rods or absent. Radula with a pair of large, curved lateral teeth, normally denticulate along the inner edge, and several smaller smooth outer teeth.

Biology: Very little is known of the biology of these animals. Some species appear to feed on flatworms living on sponges whereas others are probably generalistic carnivores.

photo © P. Armes - Tampa Bay, Florida

Gastropteron vespertilium
Gosliner & Armes, 1984

"Flapping dingbat"

Size: Up to 5 mm long

photo J. Hamann - Turks & Caicos

Gastropteron hamanni
Gosliner, 1989

"Hamann's flapping dingbat"

Size: Up to 4 mm long

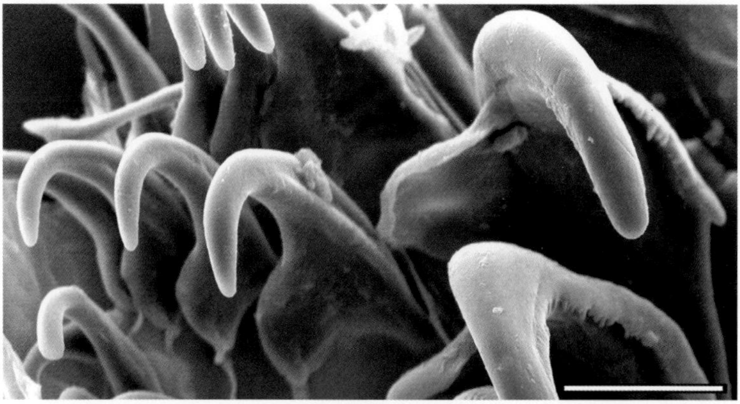

photo A. Valdés - SEM of the radula of *Gastropteron chacmol*, scale bar = 25 μm

Gastropteron vespertilium Gosliner & Armes 1984: 54-64, figs. 2-18, 23A.

Description: Background color from charcoal-gray to purplish black or black. Some specimens with a yellow, blue or green parapodial and siphonal lines.

Geographic Range: Florida.

Natural History: This species is a good swimmer; common on shallow sand in which it buries.

Gastropteron hamanni Gosliner 1989: 367-368, figs. 1H, 27A-C, 28A-B.

Description: Background color uniformly yellow; siphon with a dark brown apex. Parapodia elongate and wide, overlapping on each other when the animal is actively crawling. Head shield roughly triangular, wider anteriorly. Gill small, consisting of 2 or 3 indistinct leaflets.

Geographic Range: Turks & Caicos, Martinique.

Natural History: This species is a good swimmer; common on shallow sand in which it buries.

photo J. Hamann - Guanaja, Honduras

Gastropteron chacmol

Gosliner, 1989

"Flapping dingbat"

Size: Up to 10 mm long

photo J. Hamann - Isla Mujeres, Mexico

photo A. DuPont - mating specimens, Bahamas

photo J. Hamann - Isla Mujeres, Mexico

photo A. DuPont - trailing behavior, Bahamas

Gastropteron chacmol Gosliner 1989: 363-367, figs. 1I, 25A-E, 26A-D.

Description: Background color from deep red to black. A white or bright yellow band at the parapodial edge. Some specimens with white or yellow spots scattered over the surface of the body. Parapodia overlapping each other when the animal is in motion.

Geographic Range: Florida, Texas, Mexico, Belize, Honduras, Colombia, Venezuela, Cayman Islands, Bahamas, Brazil.

Natural History: Members of this species can actively swim and propel themselves through the water column for extended periods of time. They also follow members of the same species in a characteristic trailing or "tailgating" behavior and communicate danger using pheromones. Common on sand in which it buries.

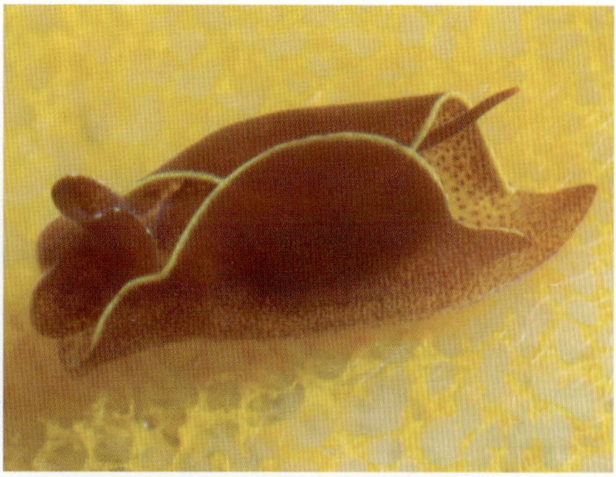

photos © T. Gosliner - Cayman Islands.

photos J. Hamann - Guanaja, Honduras

Family RUNCINIDAE H. Adams & A. Adams, 1854

Diagnosis: Body very small, elongate, smooth, narrow, normally with an elongate posterior end. Posterior end of the notum generally notched, in some species with the shell visible. Notal margin above the Hancock's organs often folded upwards forming inhalant siphons. Shell reduced or absent, cup-like or plate-like, external or internal. Gill posterior, next to the anus, composed of one or more leaves. Digestive system with gizzard plates having transverse ridges denticulate or not. Jaws composed of sharp, pointed chitinous elements. Radula with an arched or bilobed central tooth, and one or two large hooked lateral teeth on each side; smooth or denticulate.

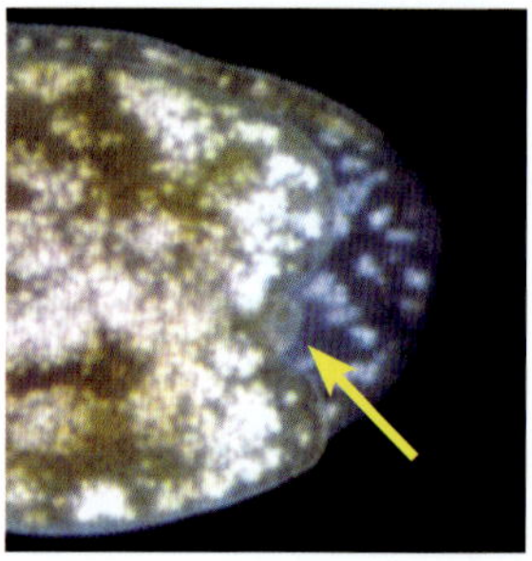

photo © L. Harris - Virgin Islands, posterior end of the body and shell (arrow)

Biology: Most runcinids are very small, worm-like sea slugs, probably the smallest species overall. They can be seasonally and locally

photo J. Hamann - Garbage Hole, Bahamas

Runcina divae
(Ev. Marcus & Er. Marcus, 1963)

Size: Up to 2 mm long

photo © T. Gosliner - Cozumel, Mexico

Runcina sp.

Size: Up to 2 mm long

extremely abundant, with population explosions that last a few days or weeks. Very little is known on their habits and food source, although they are probably herbivorous animals.

Remarks: It is likely that several Caribbean species of *Runcina* remain undescribed and some different species may be currently identified under the name *Runcina divae*. A thorough review of this group in the Caribbean is necessary to determine how many valid species are there. Some species of flat-worms are often confused with runcinids but can be distinguished by a faster moving behavior and the absence of a notch on the posterior end of the body, among other characteristics.

Ildica divae Ev. Marcus & Er. Marcus 1963: 7-8, figs. 1-7.

Description: Background color dark brown with a series of white spots, smaller on the center and larger around the mantle margin. Eyes visible, large. Anterior end of the body slightly concave. Posterior end of the body notched, with the cup-like external shell visible.

Geographic Range: Florida, Costa Rica, Curaçao, Bonaire, Bermuda, Jamaica, Bahamas, Virgin Islands, Brazil.

Description: Background color uniformly black, with no other visible pigment. Anterior end of the body slightly concave. Posterior end of the body with a short, triangular extension.

Geographic Range: Mexico, Cuba.

Remarks: This appears to be an undescribed species.

Family CYLINDROBULLIDAE Thiele, 1931

Diagnosis: Body oval with a fragile, coiled, external shell, into which the animal can retract completely. Cephalic shield shallowly furrowed medially. Operculum and parapodia absent. Female aperture below and in front of the adductor muscle, on the right side of the body. Digestive system lacking a labial cuticle. Radula uniseriate (one tooth per row), with quadrate teeth having several denticles similar in size. Ascus sack present.

Biology: Very little is known on the biology of cylindrobullids. It seems that they are probably herbivorous, possibly scraping algae growing below sediment around *Halimeda* and *Thalassia*.

Remarks: There are two Caribbean species of *Cylindrobulla*, *Cylindrobulla beauii* and *Cylindrobulla gigas*, practically indistinguishable by external characteristics or shell morphology. Examination of the radula is necessary to identify species of this genus. The systematic position of this group is subject to controversy. In recent studies it has been placed in the Cephalaspidea or in the Sacoglossa.

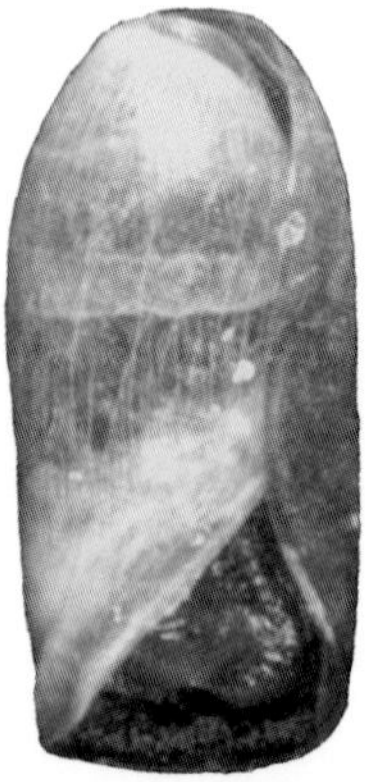

Cylindrobulla beauii
P. Fischer, 1857

"Beau's paper bubble"

Size: Up to 8 mm long

photos A. Valdés - shell, Belize; left: ventral view; right: apical view

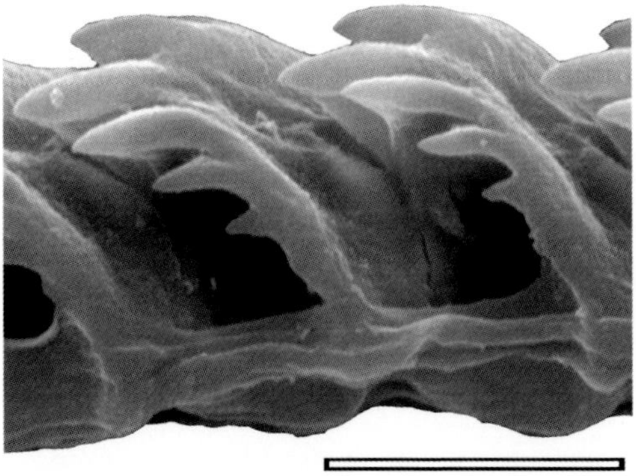

photos A. Valdés - SEM of *Cylindrobulla beauii*; left: radular teeth, scale bar = 10 µm; right: ascus, scale bar = 10 µm

Cylindrobulla beauii P. Fischer 1857: 275-276, pl. 8, figs. 8-9.

Description: Shell external, thin and fragile, cylindrical. Aperture wide anteriorly, narrowing into a sutural slit. Apex concave with the protoconch immersed.

Geographic Range: Florida, Dry Tortugas, Mexico, Belize, Colombia, Curaçao, Venezuela, Bermuda, Bahamas, Turks and Caicos, Cuba, Jamaica, Puerto Rico, Virgin Islands, Guadeloupe, Barbados, Brazil.

Natural History: Lives buried in the sediment around the roots of *Halimeda* or *Thalassia* on which it probably feeds.

Family JULIIDAE E. A. Smith, 1885

Diagnosis: Body elongate, with a pair of long auriculate, generally cylindrical rhinophores and a pair of oral lobes on the sides of the mouth. Entire body completely retractable into the shell, which is composed of two valves similar in size and shape. Shell thin or solid with the protoconch sometimes attached to it in the adult specimens. Eyes close together on a small raised prominence. Pointed posterior end of the foot, rarely projecting beyond the shell. Anus opens on the bottom of the mantle cavity, beneath and behind the gill. A strong, transverse adductor muscle at the anterior third of the visceral mass controls the closing of the shell valves. Female aperture behind the level of the adductor muscle. Radular teeth are normally finely denticulate along each side. In some species, the tip of the tooth is notched as well; others have no lateral denticles.

Biology: Juliids are tropical and temperate animals distributed throughout the world except for the eastern Atlantic and Mediterranean, where they are present only in the fossil record. They specialize in and are restricted to species of the lower intertidal and subtidal green alga *Caulerpa*. They may be locally common but extremely difficult to find because of their cryptic coloration.

Remarks: Juliids were originally discovered as fossils in European formations over a century ago, dating back to the Middle Eocene (about 49-37 million years ago). They were originally regarded as enigmatic bivalves until 1959. In that year the first living specimens were found in Japan and subsequently the group was transferred to the Sacoglossa. Because of the strict relationship between Juliids and lower intertidal and subtidal tropical ecosystems and their abundance in some fossil deposits, this group is an excellent indicator for paleo-environmental conditions.

photo J. Hamann - Bahamas

Berthelinia caribbea
Edmunds, 1963

"Caribbean bivalved snail"

Size: Up to 4 mm long

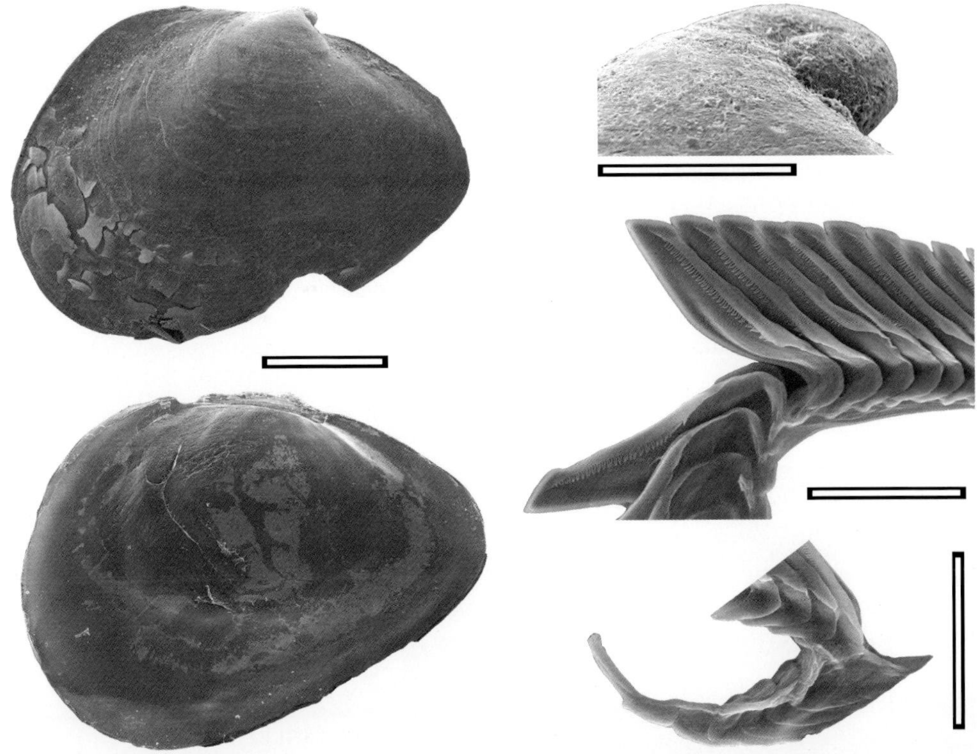

photos A. Valdés - SEM of *Berthelinia caribbea*; left: shell (upper left: left valve bearing the protoconch; lower left: right valve), scale bar = 1 mm; upper right: detail of the protoconch, scale bar = 200 µm; middle right: radular teeth, scale bar = 50 µm; lower right: ascus, scale bar = 30 µm

---

Berthelinia caribbea Edmunds 1963: 731-739, figs. 1-5.

Description: Background color translucent greenish with white patches on the head and neck and in some cases brown lines on the head and body. Shell valves translucent clear with the greenish mantle, with its distinctive pattern of brownish lines and scattered white patches, visible through the shell.

Geographic Range: Florida, Mexico, Belize, Costa Rica, Panama, Bahamas, Jamaica, Puerto Rico, Brazil.

Natural History: Feeds exclusively on *Caulerpa verticillata*.

Family VOLVATELLIDAE Pilsbry, 1895

Diagnosis: Body oval with a fragile, coiled, external shell, into which the animal can retract completely. The genus *Ascobulla* has a cephalic shield (as in cephalaspideans), with two lobes, whereas the other genus, *Volvatella* has rhinophores and oral tentacles similar to other sacoglossans. Shell with a yellowish periostracum. Anus opens on the visceral mass behind the level of the gill. Female aperture below and in front of the adductor muscle, on the right side of the body. Digestive system lacking a labial cuticle. Radula uniseriate (one tooth per row), with elongate teeth having denticles along the edges.

Biology: All volvatellids live on *Caulerpa*, and the species *Caulerpa racemosa* is the preferred food source for many species; the slugs either burrow through the sand to find and eat the rhizoids of the alga, or crawl up the stipes to feed on the vesicles.

photo J. Hamann - Harrington Sound, Bermuda

Volvatella bermudae
Clark, 1982

Size: Up to 10 mm long

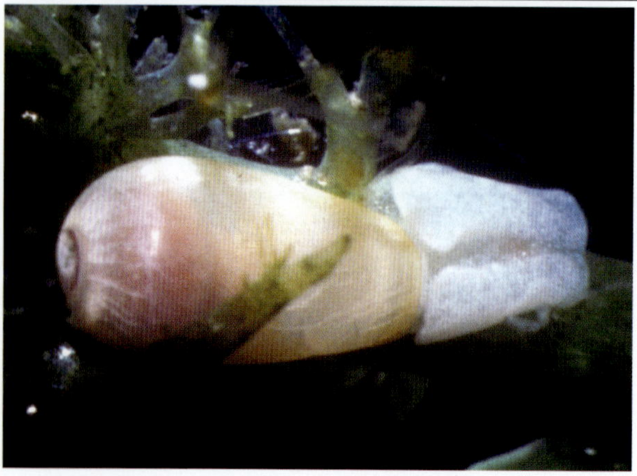

photo J. Hamann - Sebastian Inlet, Florida

Ascobulla ulla
(Er. Marcus & Ev. Marcus, 1970)

Size: Up to 8 mm long

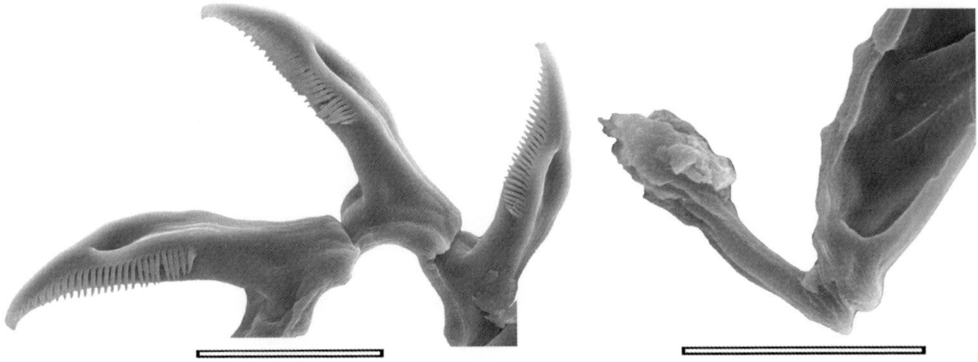

photos A. Valdés - SEM of *Ascobulla ulla*; left: radular teeth, scale bar = 30 µm; right: ascus, scale bar = 20 µm

Volvatella bermudae Clark 1982: 112-119, figs. 1-4.

Description: Background color translucent white with numerous small opaque white dots. Viscera green, visible through the shell. Shell fragile, translucent, covering the entire body.

Geographic Range: Florida, Belize, Bermuda.

Natural History: Feeds on the alga *Caulerpa racemosa*.

Cylindrobulla ulla Er. Marcus & Ev. Marcus 1970: 25-26, fig. 33.

Description: Background color translucent white with numerous small opaque white dots. Cephalic shield deeply furrowed medially. Shell fragile, translucent, covering most of the body except for the cephalic shield. Viscera visible through the shell.

Geographic Range: Florida, Mexico, Belize, Venezuela, Bermuda, Bahamas, Turks and Caicos, Cayman Islands, Virgin Islands, Brazil.

Natural History: Feeds on the alga *Caulerpa racemosa*. Animal exudes white milky fluid from the posterior end when disturbed.

Family OXYNOIDAE Stoliczka, 1868

Diagnosis: Body elongate, partially covered by a fragile external shell. Animal unable to completely withdraw inside of the shell. Shell oval, partially covered by parapodia-like extensions of the mantle in *Oxynoe*, and elongate, uncovered, surrounded by cerata-like papillae in *Lobiger*. Head with a pair of rolled rhinophores and lateral anterior prolongations. Posterior end of the body very elongate. Anus posterio-lateral in *Oxynoe* and posterior in *Lobiger*; female aperture on the right side of the body within the protection of the anterior part of the shell. Radula uniseriate, with blade-shaped teeth having fine lateral denticles.

Biology: Species of Oxynoidae are associated with algae of the genus *Caulerpa* on which they feed and remain cryptic for long periods of time. Most species display various defensive behaviors, including secretion of white chemicals, swimming, and autotomization of papillae or the posterior end of the foot.

photo A. Valdés - Puerto Vargas, Costa Rica

Lobiger souverbii
P. Fischer, 1857

"Souverbie's lobiger"

Size: Up to 30 mm long

photos J. Hamann & A. Valdés - left: specimen, Cayman Island (JH); right: head, Costa Rica (AV)

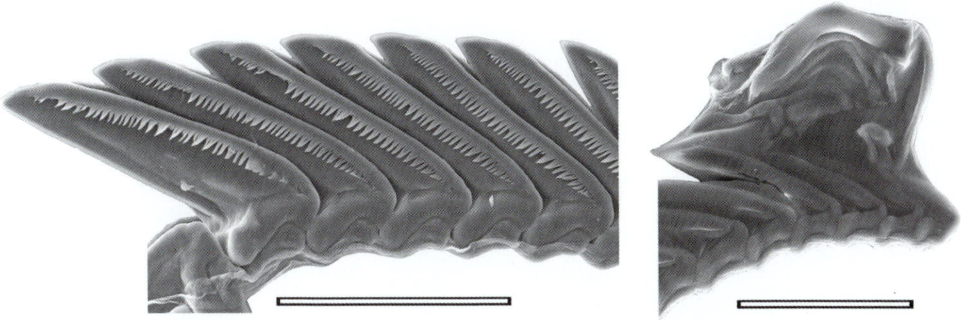

photos A. Valdés - SEM of *Oxynoe azuropunctata*; left: radular teeth, scale bar = 100 µm; right: ascus, scale bar = 50 µm

Lobiger souverbii P. Fischer 1857: 273-274, pl. 11, figs. 7-10.
Lobiger pilsbryi Schwengel 1941: 37-40, pl. 3, figs. 1-5.

Description: Background color green with dark spots irregularly distributed. Four parapodial lobes present on the sides of the body. Body, particularly the parapodial lobes, covered with conical whitish papillae. Fragile shell, green, covering the central part of the body, with thin transverse black lines and bright blue spots. Rhinophores rolled, also with conical papillae.

Geographic Range: Florida, Mexico, Honduras, Costa Rica, Venezuela, Curaçao, Cayman Islands, Jamaica, Puerto Rico, Guadeloupe, Barbados, St. Vincent & the Grenadines, Grenada, Brazil.

Natural History: Found on algae of the genus *Caulerpa* on which it feeds, lives, and lays a yellow, flat egg mass. The animal is very cryptic in its habitat. When disturbed it extends the parapodial lobes displaying the bright red areas normally hidden, it also exudes white milky fluid, and in some cases it swims away by flapping the lobes vigorously. In some cases it can autotomize the parapodial lobes.

photos J. Hamann & A. Valdés - left: specimen producing defensive secretion, Culebra, Puerto Rico (JH); right: specimen showing bright red areas, Manzanillo, Costa Rica (AV)

Oxynoe antillarum
Mörch, 1863

"Antilles oxynoe"

Size: Up to 50 mm long

photo J. Hamann - Guanaja, Honduras, dark form

photos J. Hamann - left: light form, St. Vincent & the Grenadines; right: Virgin Islands

Oxynoe azuropunctata
Jensen, 1980

"Blue spot oxynoe"

Size: Up to 20 mm long

photos J. Hamann & A. Valdés - left: Lake Surprise, Florida (JH); right: Key Largo, Florida (AV)

Oxynoe antillarum Mörch 1863: 27.
?*Oxynoe aguayoi* Jaume 1945: 22-23, pl. 2, fig. 6.

Description: Background color yellow to green, often with translucent white spots all over; some specimens with blue spots. Inner edge of the parapodia, central part of the posterior end of the foot and rhinophores normally yellowish with blue spots; from the rhinophores the yellow band extends along each side of the head. Body covered with small, conical papillae. Shell fragile, bubble-shaped, translucent, almost completely covered by the parapodia.

Geographic Range: Florida, Mexico, Belize, Honduras, Costa Rica, Panama, Venezuela, Bahamas, Curaçao, Bermuda, Cayman Islands, Jamaica, Dominican Republic, Puerto Rico, Virgin Islands, Martinique, St. Lucia, Barbados, St. Vincent & the Grenadines, Grenada, Trinidad & Tobago, Brazil.

Natural History: Feeds on species of *Caulerpa* on which it is very cryptic. When disturbed produces white secretions and can autotomize the posterior end of the body.

photos A. Valdés - Puerto Vargas, Costa Rica; left: detail of the head; right: spotted form

Oxynoe azuropunctata Jensen 1980: 283-292, figs. 1-8.

Description: Background color green, covered with large blue spots and elongate, conical, white papillae. Inner edge of the parapodia, central part of the posterior end of the foot and rhinophores with the largest papillae, whitish with blue spots. A large white patch around each eye.

Geographic Range: Florida, Belize, Bahamas, Puerto Rico, Virgin Islands.

Natural History: Feeds on species of *Caulerpa* on which it is very cryptic. When disturbed produces white secretions and can autotomize the posterior end of the body.

Family BOSELLIIDAE Ev. Marcus, 1982

Diagnosis: Body flattened, very plastic, changing from oval or almost rounded, to very elongate depending on the state of the animal. Anterior end of the foot laterally expanded into rounded corners. Rhinophores very small, grooved. Parapodia absent. Pericardium slightly raised. Radula with short to elongate, denticulate teeth. Male opening in a dorsal furrow. Anal and female apertures very close, in a ventral furrow.

Biology: Boselliids live exclusively upon green algae of the genus *Halimeda*, and their external appearance and color is very similar to that of a segment of the algae. Animals remain immobile for long periods of time on their prey, with the body longitudinally compressed. When disturbed they move greatly extending their body length.

Bosellia mimetica

Trinchese, 1891

Size: Up to 15 mm long

photos A. Valdés & J. Hamann - left: in motion, Costa Rica (AV); right: stationary, Curaçao (JH)

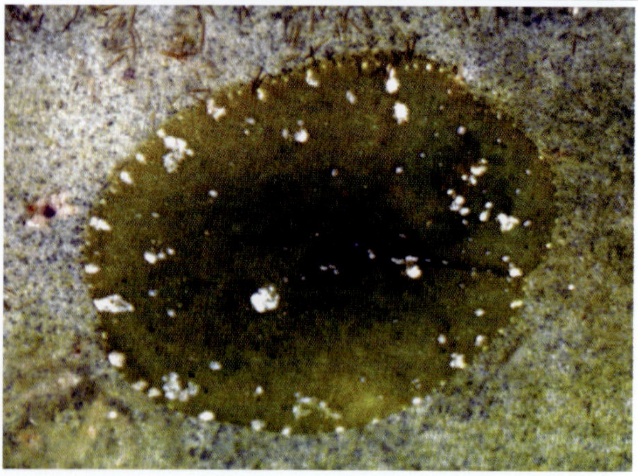

Bosellia marcusi

Ev. Marcus, 1972

Size: Up to 10 mm long

photos J. Hamann - Bahamas; left: specimen stationary; right: specimen in motion

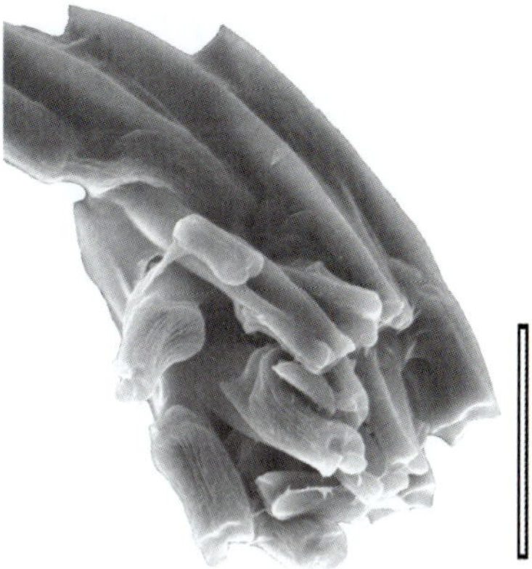

photos A. Valdés - SEM of *Bosellia mimetica*; left: radular teeth, scale bar = 30 µm; right: ascus, scale bar = 20 µm

Bosellia mimetica Trinchese 1891: 773-778, figs. 1-12.
Bosellia mimetica curasoae Er. Marcus & Ev. Marcus 1970: 50-53, figs. 91-98.

Description: Background color pale green with darker areas and opaque white spots irregularly arranged. Rhinophores short. Parapodia permanently opened.

Geographic Range: Amphiatlantic; Western Atlantic: Florida, Belize, Costa Rica, Curaçao, Bermuda, Barbados, Puerto Rico, Brazil.

Natural History: The color of the animal matches the color of the alga *Halimeda tuna* on which it lives and feeds.

Bosellia marcusi Ev. Marcus 1972: 293-295, figs. 19-22.

Description: Background color dark green with darker areas and opaque white spots irregularly arranged, sometimes covering the entire dorsum. Rhinophores short. Head set off by transverse furrows; the right one being more visible because it coincides with the position of the vagina.

Geographic Range: Florida, Costa Rica, Bahamas, Jamaica.

Natural History: Feeds on the alga *Halimeda opuntia.*

Remarks: *Bosellia corinneae* Ev. Marcus, 1973 is very similar and could constitute a synonym.

photo J. Hamann - Garbage Hole, Bahamas, white specimen

Family PLACOBRANCHIDAE Gray, 1840

Diagnosis: Body elongate, with a wing-like parapodial fold, or parapodium, along each side, folded over the dorsum joining dorsally in the center line. Foot sole transversely divided into two portions. Anterior edge of the foot rounded, sometimes expanded into angular corners, sometimes into tentaculiform corners. Parapodia simple or elaborated by the presence of folds or papillae on the edge. Rhinophores auriculate. Anus and female aperture open into a right lateral furrow; penis in a sac beneath the right rhinophore. Pericardium behind the head, usually with a number of dorsal vessels radiating from it and branching. Radular teeth variable in shape, generally pointed and denticulate along one, two or three cutting edges.

Biology: Placobranchids feed primarily on green algae, but also on diatoms. Many species sequester live chloroplasts from the algae they eat, and keep them functional in their tissue (kleptoplasty).

photo J. Hamann - Locaya, Bahamas

Thuridilla picta
(Verrill, 1901)

"Painted elysia"

Size: Up to 16 mm long

photo A. Valdés - Puerto Viejo, Costa Rica

Thuridilla mazda
Ortea & Espinosa, 2000

Size: Up to 6 mm long

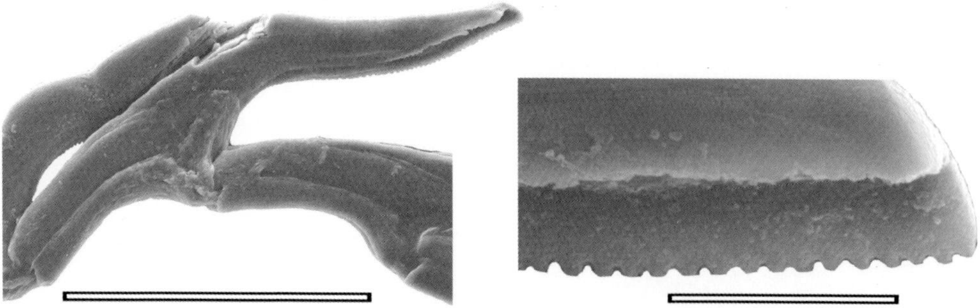

photos A. Valdés - SEM of *Elysia timida*; left: radular teeth, scale bar = 100 μm; right: detail of the denticulated cutting edge, scale bar = 10 μm

Elysia picta Verrill 1901: 30-31, pl. 4, fig. 2.
Elysia duis Ev. Marcus & Er. Marcus 1967a: 31-32, figs. 33-37.

Description: Background color yellow, with the sides of the head black with red, bright blue, and yellow spots. Rhinophores with a yellow base followed by a bright blue band, then a narrow black band, followed by a broad red area with blue or yellow spots and generally a dark apex. Parapodia yellow with a blue and a black band and a red edge with yellow spots. Lower sides of the body dark gray with yellow spots. Parapodial edge with conspicuous folds.

Geographic Range: Amphiatlantic; Western Atlantic: Florida, Bermuda, Bahamas, Curaçao, Jamaica.

Thuridilla mazda Ortea & Espinosa 2000b: 88-90, pl. 1, figs. A-F.

Description: Background color dark-brown to black, with a dorsal, opaque white area covering the center of the head and most of the rhinophores, except for the black tips. Sides of the head black with blue spots. Parapodia with light brown, bright blue, and orange half-moon-like spots and a thin white edge.

Geographic Range: Costa Rica, Bahamas, Cuba.

photo © L. Ianniello - Deerfield Beach, Florida

Thuridilla **sp.**

Size: Up to 10 mm long

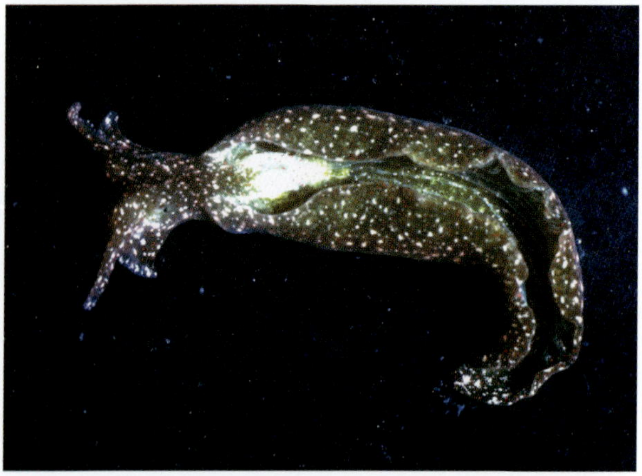

photo © T. Gosliner - Florida

Elysia chlorotica

Gould, 1870

Size: Up to 45 mm long

photo © K. Clark - Bermuda

Elysia catula

(Gould, 1870)

Size: Up to 9 mm long

Description: Background color pale green with a series of rounded yellow tubercles on the sides. Head and rhinophores mostly covered with a cream colored area with yellowish shades. Rhinophores with purple tips. Parapodial margin thick, cream with a yellow band.

Geographic Range: Florida, Colombia, Grenada.

Remarks: This is probably an undescribed species belonging to the genus *Thuridilla*, but this needs to be confirmed with anatomical studies.

Elysia chlorotica Gould 1870: 255-256, pl. 17, fi g. 251-255.

Description: Background color usually bright green, but can be reddish or grayish, depending on the amount of chlorophyll in the branches of the digestive gland that ramify throughout the body. White spots and numerous smaller red dots scattered over the body. Parapodia fold over to enclose the body.

Geographic Range: Nova Scotia, Massachusetts, Connecticut, New York, New Jersey, Maryland, Florida, Texas.

Natural History: It feeds on algae of the genus *Vaucheria*. Inhabits tidal marshes.

Placobranchus catulus Gould 1870: 256-257, pl. 17, fig. 249-250.

Description: Background color green with dense concentrations of black pigment all over the body. Whitish marks on the head, between the rhinophores, on each side of the head running across the eyes and at the edge of the parapodia. Rhinophores very short.

Geographic Range: Massachusetts, New Jersey, South Carolina, Bermuda.

Natural History: It feeds on the sea grass *Zostera*.

photo J. Hamann - Roatán, Honduras

Elysia crispata
Mörch, 1863

"Lettuce slug"

Size: Up to 150 mm long

photo A. Valdés - Key Largo, Florida

photo J. Hamann - Norman Island, Virgin Islands

photo J. Hamann - juvenile specimen, Pt. Antonio, Jamaica

Elysia ornata
(Swainson, 1840)

"Ornate elysia"

Size: Up to 55 mm long

photos J. Hamann - left: adult specimen, Terre d'en Haut, Guadeloupe; right: head, Aruba

Elysia (Tridachia) crispata Mörch 1863: 40-41.
Tridachia schrammi Mörch 1863: 41.
Elysia verrilli Pruvot-Fol 1946: 39 [non *Elysia verrilli* Thiele, 1931].
Elysia (Elysiopterus) pruvotfolae Er. Marcus 1957: 415.
Tridachia whiteae Er. Marcus 1957: 416.
Elysia clarki Pierce *et al.* 2006: 26-36, figs. 1B, 1D, 2, 4A, 5A, 5C, 5E, 5G, 6A-B, 7.

Description: Background color variable, from cream to green, with lighter oval spots all over the body. Parapodia greatly enlarged and folded. Edge of the parapodia with a distinctive color, variable from cream, to brown, red, green or blue.

Geographic Range: Florida, Dry Tortugas, Mexico, Belize, Honduras, Costa Rica, Colombia, Venezuela, Aruba, Curaçao, Bonaire, Venezuela, Bermuda, Cayman Islands, Bahamas, Jamaica, Haiti, Puerto Rico, Virgin Islands, St. Maarten/St. Martin, Antigua, St. Lucia, Martinique, Guadeloupe, Turks & Caicos, St. Vincent & the Grenadines, Barbados, Trinidad & Tobago.

Natural History: Feeds on the green algae *Caulerpa sertularioides*.

Remarks: *Elysia clarki* was described for the Florida Keys populations without comparison to other Caribbean populations except for specimens from the Virgin Islands. Morphological differences between Florida and Virgin Islands populations fit within the intraspecific variability of *E. crispata*.

photos J. Hamann - left: specimen with open parapodia, Abaco, Bahamas; right: lateral view of adult specimen, Aruba

---

Thallepus ornatus Swainson 1840: 251, 359.
Pterogasteron marginatum Pease 1871: 304, pl. 21, fig. 3.

Description: Background color translucent green with a black parapodial margin and a submarginal yellow or orange band. Rhinophores orange, with a white base and a black tip. Numerous black and opaque white dots all over the body.

Geographic Range: Circumtropical; Western Atlantic: Florida, Belize, Honduras, Costa Rica, Colombia, Venezuela, Bahamas, Curaçao, Bermuda, Jamaica, Puerto Rico, Virgin Islands, Martinique, Turks & Caicos, Barbados, St. Vincent & the Grenadines, Grenada, Trinidad & Tobago, Brazil.

Natural History: Feeds on algae of the genus *Bryopsis*.

Elysia papillosa
Verrill, 1901

Size: Up to 35 mm long

photo J. Hamann - left: Chimana Grande, Venezuela; right: view of hepatic vessels, St. Lucia

Elysia canguzua
Er. Marcus, 1955

Size: Up to 40 mm long

photo A. Valdés - Puerto Vargas, Costa Rica

Elysia evelinae
Er. Marcus, 1957

Size: Up to 8 mm long

photo © A. Bernecker - Puerto Vargas, Costa Rica

Elysia papillosa Verrill 1901: 31, pl. 4, fig. 3.

Description: Background color pale green to greenish gray, with a broad whitish, yellowish or brownish band along the parapodial edge. A large white blotch inside of the parapodia, on the right side of the pericardium. Body with numerous, densely arranged, conspicuous papillae, usually white, covering most parts of the animal, except the foot.

Geographic Range: Florida, Mexico, Belize, Honduras, Costa Rica, Panama, Venezuela, Cayman Islands, Curaçao, Bahamas, St. Lucia, Guadeloupe, Martinique, Trinidad & Tobago, Grenada, Bermuda.

Natural History: Feeds on green algae of the genera *Halimeda*, *Penicillus* and *Udotea*.

Remarks: The identity of this species is unclear, it was described based on specimens from Bermuda. The specimens illustrated here could belong to an undescribed species.

Elysia canguzua Er. Marcus 1955: 111-113, fig. 45-48, 60-65.
Elysia eugeniae Ortea & Espinosa 2002: 130-133, figs. 1A-C, 2A-D; pl. 1, fig. A.

Description: Background color green, produced by numerous small green branches. Body covered with bright red-orange dots, a few bright blue spots and some black pigment. Anus conspicuous, situated dorsally on the right hand side at the beginning of the right parapodium. Parapodia open near the anterior end forming two wing-like folds.

Geographic Range: Costa Rica, Brazil.

Natural History: Feeds on *Codium*.

Remarks: The background green color composed of numerous branches, the orange and bright blue spots and the conspicuous anus situated on the dorsal right side of the body are characteristic of this species noted on the original descriptions of *Elysia canguzua* and *E. eugeniae.*

Elysia evelinae Er. Marcus 1957: 410-416, figs. 48-57.

Description: Background variable color from pale cream to brownish gray, with numerous red and opaque white spots all over the body. A conspicuous black patch around the anal papillae, situated on the right side, at the base of the head.

Geographic Range: Florida, Costa Rica, Brazil.

Natural History: Feeds on diatoms of the genus *Biddulphia*.

Remarks: The original description of this species states that the anus is surrounded by a black ring, which matches the specimen illustrated here.

Elysia tuca
Ev. Marcus & Er. Marcus, 1967

Size: Up to 40 mm long

photos J. Hamann - left: Isla Mujeres, Mexico; right: atypical specimen, Curaçao

Elysia subornata
Verrill, 1901

"Dark-rimmed elysia"

Size: Up to 70 mm long

photo J. Hamann - Flamingo Bay, Virgin Islands

photos J. Hamann - left: Isla Mujeres, Mexico; right: view of the renopericardial lump, Aruba

Elysia tuca Ev. Marcus & Er. Marcus 1967a: 29-31, figs. 28-32.

Description: Background color from light to dark green or brown. An opaque white patch on the head behind the rhinophores and a white line from the patch running up the basal half of each rhinophore. Rhinophores with a white upper region, sometimes the upper half, while the lower half is often mottled with green or brown. Prominent white rounded papillae on the rhinophores and scattered on the parapodia. Parapodia normally joined on top, leaving two open areas, one in the middle of the body and the other posteriorly.

Geographic Range: Florida, Mexico, Honduras, Costa Rica, Panama, Colombia, Bermuda, Curaçao, Cayman Islands, Jamaica, Puerto Rico, Virgin Islands, St. Maarten/St. Martin, St. Lucia, Barbados, St. Vincent & the Grenadines, Grenada, Brazil.

Natural History: Usually found on species of alga *Halimeda*, which constitutes its food source. Largest specimens found during reproduction on *Caulerpa*.

Elysia subornata Verrill 1901: 29-30, pl. 4, fig. 4.

Description: Background color highly variable, ranging from yellow green to dusky olive, to reddish, usually with a brownish to black thin line on the parapodial margins. Head normally with a white patch entering the rhinophores. Parapodia large, partially overlapping in the living animals, covering portions of each other, with thickened, irregular margins. Body covered with white papillae of different sizes and in varying densities. Rhinophores papillate, with bands of white, gray and or brown. Renopericardial lump extremely elongate.

Geographic Range: Florida, Mexico, Belize, Bermuda, Bahamas, Aruba, Cayman Islands, Jamaica, Puerto Rico, Virgin Islands, Martinique, Grenada, Trinidad & Tobago.

Natural History: Feeds on the green algae *Penicillus dumetosus* and *Udotea flabellum*.

photos J. Hamann - Belize; left: papillate specimen: right: color variation among individuals

Elysia timida
(Risso, 1818)

Size: Up to 20 mm long

photos A. DuPont & J. Hamann - left: Bahamas (AD); right: Cayman Islands (JH)

Elysia serca
Er. Marcus, 1955

"Sea grass elysia"

Size: Up to 8 mm long

photos J. Hamann & A. DuPont - left: Indian River, Florida (JH); right: Bahamas (AD)

Elysia flava
Verrill, 1901

Size: Up to 20 mm long

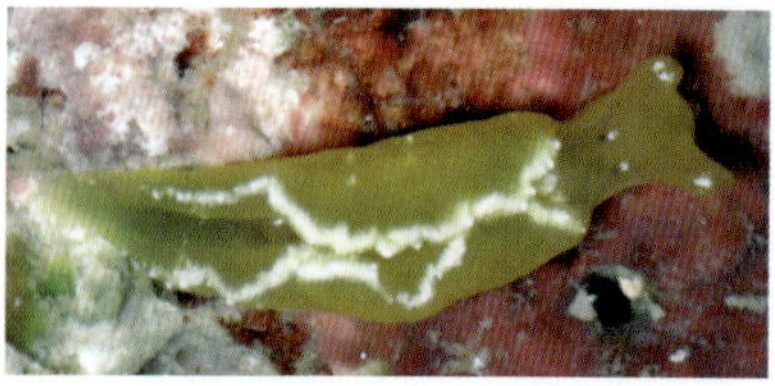

photos J. Hamann & A. Valdés - left: Chimana Grande, Venezuela (JH); right: Costa Rica (AV)

Notarchus timidus Risso 1818: 376.
Elysia viridis lactea Bergh 1880b: 3.
Elysia cornigera Nuttall 1989: 302-306, figs. 1-8.

Description: Background color opaque white with numerous red dots. Body covered with irregular papillae. Parapodia medially forming a characteristic tall lobe.

Geographic Range: Amphiatlantic; Western Atlantic: Florida, Cuba, Cayman Islands, Bahamas.

Natural History: Known to feed on algae of the genera *Padina* and *Acetabularia*.

Remarks: The western Atlantic population was named *Elysia cornigera* until recently synonymized with the eastern Atlantic *Elysia timida*. There are consistent external differences between western and eastern populations and their status needs to be clarified using anatomical data.

Elysia serca Er. Marcus 1955: 113-115, fig. 49-52, 59.
Elysia clena Er. Marcus & Ev. Marcus 1970: 49, figs. 81, 89-90.

Description: Background color ranges from dark brown, to light brown, or green. Head proportionally large; parapodia reduced. A characteristic pattern of three whitish patches, one in front of the heart, covering the head, and one midway along the edge of each of the parapodia.

Geographic Range: Florida, Curaçao, Barbados, Brazil.

Natural History: Feeds on sea grasses of the genera *Halophila*, *Halodule* and *Thalassia*.

Elysia flava Verrill 1901: 30, pl. 4, fig. 1.

Description: Background color light yellow with some small, opaque white papillae. Parapodia yellow to light green with an opaque white edge bearing numerous papillae. Inner and outer sides of the parapodia sometimes with brownish black pigment.

Geographic Range: Amphiatlantic; Western Atlantic: Honduras, Costa Rica, Panama, Venezuela, Bermuda, Bahamas, Jamaica, Puerto Rico, St. Lucia, Guadeloupe, St. Vincent & the Grenadines, Grenada, Brazil.

Natural History: Normally found under rocks.

Elysia nisbeti
Thompson, 1977

Size: Up to 10 mm long

photo J. Hamann - Bermuda

Elysia pratensis
Ortea & Espinosa, 1996

Size: Up to 30 mm long

photos J. Hamann - left: adult specimen, Roatán, Honduras; right: egg mass, Bird Cay, Belize

Elysia zuleicae
Ortea & Espinosa, 2002

Size: Up to 17 mm long

photos A. Bernecker & J. Hamann - left: Puerto Vargas, Costa Rica (AB); right: Jamaica (JH)

Elysia nisbeti Thompson 1977: 126, figs. 25B-C, 26d.

Description: Background color light green, transitioning to the pale brown of the parapodial margin. Body and rhinophores covered with rows of white papillae. A "Y" brown pattern on the head enters the rhinophores. Body very elongate, tapering into a sharp posterior end; parapodia narrow.

Geographic Range: Jamaica, Bermuda.

Elysia pratensis Ortea & Espinosa 1996: 116-119, figs. 1-3.

Description: Background color ranging from green to yellowish green, with a pattern of white longitudinal lines all over the body and parapodia. Sometimes lines accentuated by the presence of alternating brown lines. Unbranched, whitish papillae scattered over the parapodia. Inside of the parapodia dark green. A black line along the parapodial margin.

Geographic Range: Florida, Mexico, Honduras, Belize, Bahamas.

Natural History: Found on seagrass beds feeding on *Penicillus*.

Elysia zuleicae Ortea & Espinosa 2002: 133-139, figs. 3A-C, 4A-E, 5A-C, 6A-D, 7A-C, pl. 1, fig. B.

Description: Background color variable from pale to dark green with numerous irregular white patches, more densely packed on the head. Juveniles much lighter. Posterior end of the foot dark gray or black, very elongate. Body and rhinophores covered with papillae.

Geographic Range: Cuba, Costa Rica, Jamaica.

Elysia patina

Ev. Marcus, 1980

Size: Up to 20 mm long

photos J. Hamann - left: Bat Cave Point, Martinique; right: Isla Mujeres, Mexico

Elysia scops

Ev. Marcus & Er. Marcus, 1967

Size: Up to 20 mm long

photos J. Hamann - left: Guanaja, Honduras; right: view of hepatic vessels, Turks & Caicos

Elysia sp. 1

Size: Up to 12 mm long

photos J. Hamann - Cozumel, Mexico; left: animal in motion; right: animal stationary

Elysia patina Ev. Marcus 1980: 72-73, figs. 23-24, 36, 41-43, 57.

Description: Background color from yellowish gray to dark green. Head with abundant white pigment, entering the rhinophores. Rhinophores very elongate, normally with a brown band at mid-length. Body and rhinophores covered with papillae. Parapodia narrow, with a thickened, irregular margin. Pericardium small, with numerous branched vessels emerging from it; two large glands on the posterior, inner side of the rhinophores.

Geographic Range: Florida, Mexico, Costa Rica, Bahamas, Martinique, St. Vincent & the Grenadines.

Natural History: Probably feeds on *Udotea*.

Elysia cauze scops Ev. Marcus & Er. Marcus 1967a: 28-29, figs. 26-27.

Description: Background color yellowish to brownish, with white pigmentation on the dorsum and parapodia. Rhinophores dark brown, covered with large white papillae. Parapodia wide, with irregular, thickened edges. Pericardium short with branched vessels.

Geographic Range: Florida, Honduras.

Remarks: This species has been regarded as a synonym of *Elysia subornata*, but it differs from it by having a shorter body with larger eyes and a much smaller renopericardial lump.

Description: Background color yellowish green with reddish dots and opaque white patches. Parapodial margins white with some black spots. Rhinophores with some brownish red and opaque white pigment. Some red spots around the anus.

Geographic Range: Mexico.

Natural History: Feeds on the green alga *Penicillus*.

Remarks: This appears to be an undescribed species.

Elysia sp. 2

Size: Up to 19 mm long

photos J. Hamann - Cozumel, Mexico; left: specimen in motion; right: vessels and egg mass

Elysia sp. 3

Size: Up to 8 mm long

photos J. Hamann - Cozumel, Mexico; left: dorsal view; right: lateral view

Elysia sp. 4

Size: Up to 17 mm long

photos J. Hamann - left: Soutiere Bay, St. Lucia; right: Rendezvous Cay, Belize

Description: Background color uniform dark green with numerous, minute white and black spots all over the body. Parapodia narrow with numerous opaque white marginal glands. Rhinophores with white pigment and some papillae. Renopericardial lump very elongate.

Geographic Range: Mexico.

Remarks: This appears to be an undescribed species.

Natural History: Found on *Avrainvillea*.

Description: Background color uniform reddish to greenish with a few opaque white spots on the head and parapodial border. Parapodia narrow with a black marginal line. Rhinophores with a few white papillae.

Geographic Range: Mexico

Remarks: This appears to be an undescribed species.

Description: Background color pale green with minute reddish dots and some white pigment. Parapodia very narrow with thickened edges. Rhinophores short, with some inconspicuous white papillae and a few red dots.

Geographic Range: Belize, St. Lucia.

Natural History: This species is able to swim when disturbed.

Remarks: This appears to be an undescribed species.

Elysia sp. 5

Size: Up to 40 mm long

photo J. Hamann - True Blue Point, Grenada

Elysia sp. 6

Size: Up to 15 mm long

photos © P. Armes - Florida; left: two specimens; right: specimen showing dorsal vessels

Elysia sp. 7

Size: Up to 15 mm long

photo J. Hamann - Isla Mujeres, Mexico

Description: Background color pale green with minute brown dots. Body covered with conical papillae. Parapodia very wide with a narrow golden line on the edges. Rhinophores long, covered with papillae.

Geographic Range: Grenada.

Natural History: This species leaves very characteristic grazing circular marks on the *Caulerpa prolifera* they eat.

Remarks: This appears to be an undescribed species.

photo J. Hamann - egg mass, Grenada

Description: Background color ochre-brown with numerous opaque white papillae. Parapodia with a black rim on the edge. Rhinophores long, white, with some inconspicuous papillae and a few brown dots. Dorsum with the reproductive and digestive glandular tissue visible through the skin.

Geographic Range: Florida.

Remarks: This appears to be an undescribed species.

Description: Background color translucent white with large areas of opaque white pigment and a few brown spots. Body covered with low, white papillae. Parapodia thickened. Rhinophores short, with some inconspicuous white papillae.

Geographic Range: Mexico.

Remarks: This appears to be an undescribed species.

Family LIMAPONTIIDAE Gray, 1847

Diagnosis: Body oval to elongate. Anterior edge of the foot rounded or with tentaculiform corners; in *Costasiella* mouth in a deep furrow between two ridges. Rhinophores long and fusiform, reduced, or absent; sometimes with a lateral groove at the base. Eyes behind the rhinophores, usually in clear patches. Cerata from globular to slender, absent in some species and varying considerably in shape in others, with branches of the digestive and sometimes of the albumen gland. Anus opens on the front edge of the pericardium or on its front right side. Male aperture below and behind the right rhinophore; one or two female apertures more posteriorly. A raised vaginal ridge below the anterior cerata indicates that the vaginal bursa lies just below the skin and insemination is hypodermic. Radular teeth non-denticulate, blade-like, wooden shoe-shaped with a spoon like cavity that fits over the base of the next tooth, or awl-shaped with a sharp point.

Biology: Limapontiids feed and live on several green algae, although some species feed on opisthobranch eggs. Several species vary considerably in color, from black to green forms, depending on the depth at which they are found. A number of limapontiids can tolerate reduced salinities.

Costasiella ocellifera
(Simroth, 1895)

"Eyespot costasiella"

Size: Up to 10 mm long

photos J. Hamann - left: Lark Cay, Belize; right: juvenile specimens, Pt. Antonio, Jamaica

Costasiella nonatoi
Ev. Marcus & Er. Marcus, 1960

Size: Up to 5 mm long

photos A. Valdés & J. Hamann - left: Manzanillo, Costa Rica (AV); right: Trujillo, Honduras (JH)

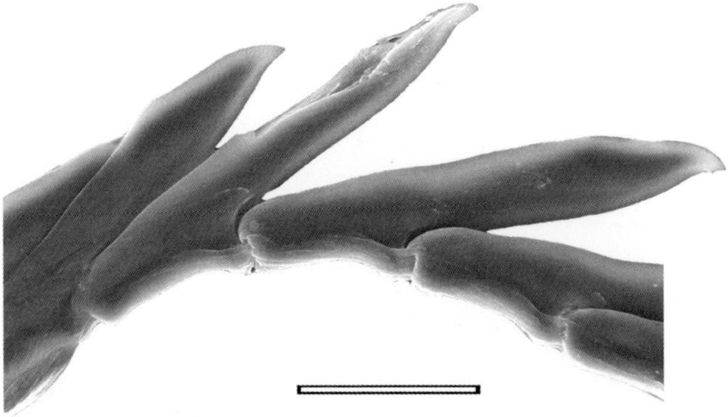

photo A. Valdés - SEM of the radular teeth of *Costasiella ocellifera*, scale bar = 25 µm

Doto ocellifera Simroth 1895: 168-170, pl. 20, figs. 6-10.
Stiliger lilianae Ev. Marcus & Er. Marcus 1969: 7-12, figs. 22-28.

Description: Background color translucent clear, covered with black and white spots. Head with orange-yellow patches, a characteristic blue spot between the rhinophores, and two conspicuous, large black eyes situated close together. Cerata inflated, narrowing into a long thin tapering tip. Ceratal wall translucent, almost transparent, with black and white spots, green branches of the digestive gland, a wide orange-yellow subapical band, and translucent tips.

Geographic Range: Florida, Mexico, Belize, Honduras, Costa Rica, Bermuda, Bahamas, Cayman Islands, Puerto Rico, Jamaica, St. Maarten/St. Martin, St. Lucia, Martinique, St. Vincent & the Grenadines, Grenada, Brazil.

Natural History: Lives and feeds on the surface of the green alga *Avrainvillea longicaulis.*

Costasiella nonatoi Ev. Marcus & Er. Marcus 1960: 149-151, figs. 26-33.

Description: Background color black with opaque white spots, white pigment along the anterior edge of the foot, and a white line up the outer edge of the rhinophores to the white tip. A large translucent white patch around the eyes. Eyes conspicuous, close to each other. Cerata black with numerous opaque white spots increasing in density at tip.

Geographic Range: North Carolina, Florida, Costa Rica, Cayman Islands, Puerto Rico, Bahamas, Bermuda, Brazil.

Natural History: Burrows into the thalus of the green alga *Avrainvillea nigricans* to feed.

Ercolania viridis
A. Costa, 1866

Size: Up to 10 mm long

photo J. Hamann - Discovery Bay, Jamaica

photo A. Valdés - Captiva Island, Florida, dark specimen

photo © L. Harris - juvenile specimen, Guana Island, Virgin Islands

Ercolania fuscata
(Gould, 1870)

"Dusky stiliger"

Size: Up to 7.5 mm long

photos J. Hamann & A. Valdés - left: Indian River, Florida (JH); right: Captiva, Florida (AV)

Ercolania viridis A. Costa 1866: 75, pl. 3, figs. 1-2.
Embletonia funerea A. Costa 1867a: 36, pl. 2, fig. 5.
Ercolania selva Ortea & Espinosa 2001b: 45-47, figs. 11A-D.

Description: Background color variable from translucent gray to black, with a pink spot at the base of the head. Translucent animals with green digestive ducts showing through; digestive branches enter the head up to the base of the rhinophores and the cerata. Rhinophores and cerata with numerous opaque white spots and sometimes pinkish pigment. Cerata very plastic, changing shapes from globose to elongate depending on the state of the animal, arranged in one or two rows on the sides of the body leaving the center of the dorsum visible.

Geographic Range: North Carolina, Florida, Costa Rica, Cayman Islands, Puerto Rico, Jamaica, Virgin Islands, Bermuda, Brazil.

Natural History: Feeds on the green algae *Chaetomorpha*, *Cladophora* and *Cladophoropsis*.

Remarks: This species shows extreme color and ceratal shape variation.

photos J. Hamann - Harrington Sound, Bermuda

Calliopaea (?) fuscata Gould 1870: 250, pl. 16, fig. 218-221.
Stiliger vanellus Er. Marcus 1957: 402-405, figs. 30-34.

Description: Background color translucent gray, with numerous darker spots, sometimes in a transverse pattern, varying in color and density between specimens. Cerata oval, translucent with dark and white spots, starting about one-third of the way down the body. Digestive gland branches visible in some specimens in the cerata and body.

Geographic Range: Nova Scotia, New Hampshire, Massachusetts, New York, New Jersey, Virginia, Florida, Curaçao, Jamaica, Puerto Rico, Brazil.

Natural History: Feeds on filamentous algae of the genera *Cladophora* and *Chaetomorpha*.

photo J. Hamann - Discovery Bay, Jamaica

Ercolania caerulea
Trinchese, 1872

Size: Up to 16 mm long

photo J. Hamann - St. Lucia

photo © L. Harris - ventral view, Guana Island, Virgin Islands

photo J. Hamann - Discovery Bay, Jamaica

Placida cf. *dendritica*
(Alder & Hancock, 1843)

Size: Up to 11 mm long

Ercolania caerulea Trinchese 1872: 86-132, pls. 4-12.
Stiliger cricetus Er. Marcus & Ev. Marcus 1970: 40-41, figs. 74-76.

Description: Background color translucent almost transparent, with the green digestive gland ducts showing through; digestive branches enter well into the head, rhinophores, and cerata. Rhinophores and cerata with numerous opaque white spots. Numerous cerata covering the dorsum. Cerata short, globose, tapering into an elongate apex with numerous white spots, giving the tips a white appearance.

Geographic Range: Florida, Venezuela, Curaçao, Jamaica, Virgin Islands, St. Kitts, St. Lucia, Brazil.

Natural History: Found inside clumps of the green bubble algae *Dictyosphaeria cavernosa* and *Valonia* sp.

photos © L. Harris & © K. Clark - left: juvenile specimen, Guana Island, Virgin Islands (LH); right: adult specimen next to the egg mass, Florida (KC)

Calliopaea dendritica Alder & Hancock 1845-55[1848]: fam. 3, pl. 40, figs. 1-7.
[several other synonyms not listed]

Description: Background color translucent almost transparent with some opaque white spots and the green digestive gland ducts showing through; digestive branches enter well into the head, rhinophores, and cerata. Cerata on the sides, leaving the center of the dorsum visible; cerata translucent, oval, with the digestive gland highly ramified.

Geographic Range: Cosmopolitan; Western Atlantic: North Carolina, Costa Rica, Curaçao, Jamaica.

Remarks: *Placida dendritica* may constitute a complex of different species. The specimen illustrated here lacks lateral ceratal branches and could constitute an undescribed species.

photo A. Valdés - Puerto Vargas, Costa Rica

photo J. Hamann - Bermuda

Placida kingstoni

Thompson, 1977

Size: Up to 15 mm long

photo J. Hamann - Pt. Antonio, Jamaica

Placida verticilata

Ortea, 1981

Size: Up to 4 mm long

Placida kingstoni Thompson 1977: 138-139, figs. 26I, 32B-E.

Description: Background color translucent almost transparent, with the green digestive gland ducts showing through; digestive branches enter well into the head, rhinophores, and cerata. Pericardial area opaque white, very conspicuous. Cerata only on the sides, leaving the center of the dorsum visible; cerata translucent, very elongate and narrow, with the digestive gland highly ramified.

Geographic Range: Florida, Costa Rica, Jamaica, Martinique, Bermuda.

Natural History: Feeds on algae of the genera *Bryopsis* and *Cladophora*.

photos J. Hamann & A. Valdés - left: Martinique (JH); right: ventral view of head, Costa Rica (AV)

Placida verticilata Ortea 1981: 191-194, fig. 10, 11, 12B.

Description: Background color translucent almost transparent with numerous opaque white spots and the green digestive gland ducts showing through; digestive branches enter well into the head, rhinophores, and cerata. Cerata on the sides, leaving the center of the dorsum visible; cerata translucent, elongate, with numerous opaque white spots more densely arranged on the tips and the digestive gland highly ramified.

Geographic Range: Amphiatlantic; Western Atlantic: Florida, Costa Rica, Jamaica.

Family HERMAEIDAE H. Adams & A. Adams, 1854

Diagnosis: Body elongate, narrow, with auriculate rhinophores, flared and stepped at the tips. Eyes behind the rhinophores, normally surrounded by a white or colorless area. Anterior edge of the foot rounded or with two oral lobes projecting laterally above the mouth. Anus opens in the body mid-line. Male reproductive aperture on the right side behind and below the rhinophore, a short distance anterior to the female aperture. Cerata vary from elongate to fusiform, sometimes irregular, with the visible branches of the digestive gland. Radular teeth short with blunt, denticulate cusps, or sabot shaped.

Biology: Hermaeids feed on a large range of green and red algae.

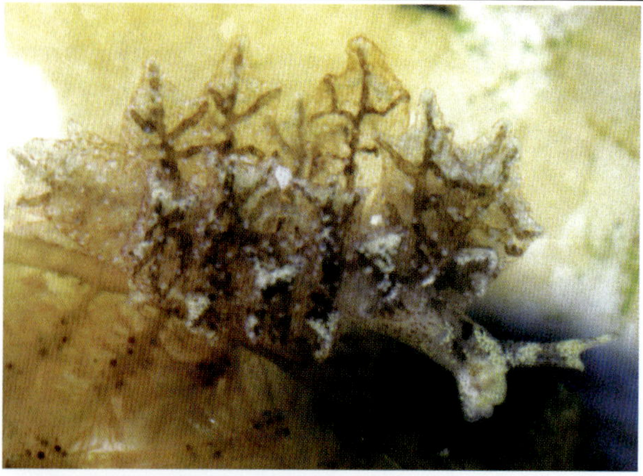

photo J. Hamann - Store Bay, Tobago

Hermaea bifida
(Montagu, 1815)

Size: Up to 5 mm long

Hermaea cruciata
Gould, 1870

Size: Up to 5 mm long

photos © K. Clark & © T. Gosliner - Florida

photos A. Valdés - SEM of *Hermaea cruciata*; left: radular teeth, scale bar = 30 µm; right: ascus, scale bar = 30 µm

Doris bifida Montagu 1815: 198, pl. 14, fig. 2.

Description: Background color translucent, almost transparent, with numerous opaque white, yellow and brown spots all over the body and the brown digestive system showing through. Rhinophores translucent with numerous yellow spots. Cerata translucent, elongate, with an irregular edge and numerous opaque white spots; the digestive gland branches visible and highly ramified.

Geographic Range: Amphiatlantic; Western Atlantic: Costa Rica, Tobago.

Hermaea cruciata Gould 1870: 253-254, pl. 17, fig. 256.
Hermaea coirala Er. Marcus 1955: 108-111, fig. 38-44.

Description: Background color translucent, almost transparent, with numerous opaque white spots and the brown digestive system showing through. Cerata translucent, elongate, with numerous opaque white spots and the digestive gland branches visible; digestive branches very narrow widening suddenly near the apex and then narrowing again to a blunt tip, forming a cross-like design.

Geographic Range: Massachusetts, New York, Florida, Costa Rica, Trinidad & Tobago, Brazil.

Natural History: Feeds on filamentous red algae.

Family CALIPHYLLIDAE Tiberi, 1881

Diagnosis: Body oval, broad, usually rounded in front and tapering behind to a blunt point. Rhinophores bifid, in some species for only a short part of their length. Oral tentacles rounded or pointed, hollowed below. In *Cyerce* foot sole divided at the anterior third by a transverse groove. Eyes visible behind the base of the rhinophores. Cerata variable in shape, from flat and leaf-like to inflated. Anus opens on a papilla to the right of the centre line midway between the eyes and pericardium or in a right lateral position below or between the lowest cerata. Reproductive apertures open below and behind the right rhinophore. Albumen gland absent from the cerata, digestive gland absent in *Cyerce*, present in most other genera. Edges of the cerata smooth, or decorated with defensive and adhesive glands. Radular teeth pointed and smooth, short and laterally denticulate, or very long and coarsely denticulate.

Biology: Caliphyllidiids feed on a large range of green algae including *Bryopsis*, *Penicillus*, *Halimeda*, *Chlorodesmis*, *Valonia* and *Cladophora*. Most species autotomize the cerata when disturbed.

Cyerce cristallina
(Trinchese, 1881)

"Harlequin glass slug"

Size: Up to 50 mm long

photos J. Hamann - left: Crocus Bay, Anguilla; right: San Andrés, Colombia.

Cyerce antillensis
Engel, 1927

"Antillean glass slug"

Size: Up to 60 mm long

photos A. Valdés & J. Hamann - left: Puerto Vargas, Costa Rica (AV); right: Bahamas (JH)

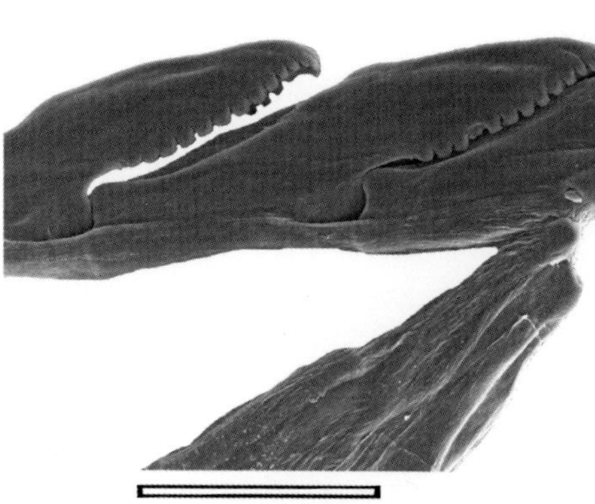

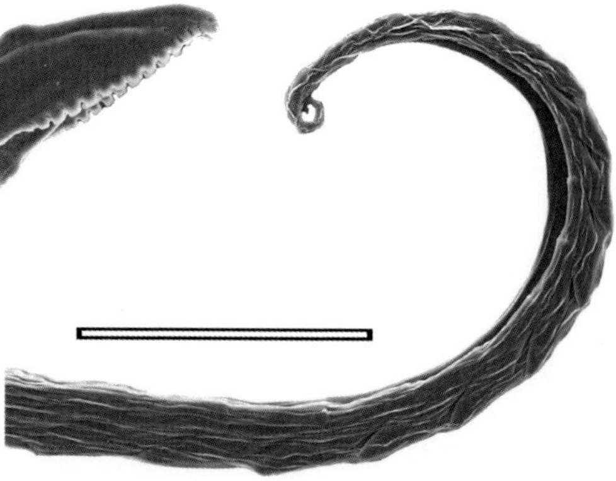

photos A. Valdés - SEM of *Polybranchia viridis*; left: radular teeth, scale bar = 100 μm; right: ascus, scale bar = 200 μm

Lobiancoia cristallina Trinchese 1881: 116.
Cyerce iheringi Pelseneer 1892: 19.

Description: Background color translucent white with distinct reddish gray markings on the head and rhinophores. Oral tentacles translucent. Cerata leaf-like, numerous, translucent, often edged and spotted with white, with a burgundy or reddish gray marking down the mid-line, darker on the tips.

Geographic Range: Amphiatlantic; Western Atlantic: Florida, Colombia, Curaçao, Bermuda, Bahamas, Jamaica, St. Maarten/St. Martin.

Natural History: It can autotomize the cerata when disturbed.

Cyerce antillensis Engel 1927: 117, fig. 1.
Cyerce habanensis Ortea & Templado 1989: 11-14, figs. 1-2.

Description: Background color translucent, with greenish, sometimes yellowish or white, viscera showing through. Opaque white pigmentation on the dorsum and sometimes a brownish pattern on the head. Pericardium opaque white. Cerata inflated, translucent, almost transparent with some scattered white spots; upper edge irregular with numerous opaque white glands. Center of the cerata with no branches of the digestive gland, but with yellowish granules and normally a large, dark spot.

Geographic Range: Florida, Mexico, Belize, Honduras, Costa Rica, Curaçao, Bermuda, Cayman Islands, Cuba, Bahamas, Jamaica, Puerto Rico, Virgin Islands, Barbados, Tobago.

Natural History: Feeds on algae of the genera *Penicillus*, *Udotea* and *Halimeda*.

Cyerce edmundsi
Thompson, 1977

Size: Up to 10 mm long

photos J. Hamann - Roatán, Honduras; left: dorsal view; right: lateral view

Polybranchia viridis
(Deshayes, 1857)

Size: Up to 70 mm long

photo J. Hamann - Daaiboui Baai, Curaçao

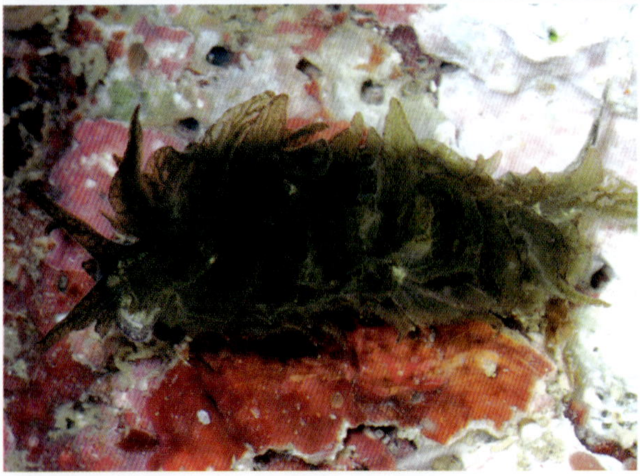

Caliphylla mediterranea
A. Costa, 1867

Size: Up to 20 mm long

photo A. Valdés - Puerto Vargas, Costa Rica

Cyerce edmundsi Thompson 1977: 136-137, figs. 22I-J, 24D.

Description: Background color whitish gray, with some brown pigment on the dorsum and cerata. Opaque white spots all over the body. Eyes large, close together. Oral tentacles elongate. Cerata short, flattened, irregularly edged, with large opaque white spots and some black dots.

Geographic Range: Honduras, Jamaica.

Hermaea viridis Deshayes 1857: 141.

Description: Background color translucent, almost transparent, with the viscera visible, giving the animal a green to pale gold tinge. Cerata and rhinophores covered with small papillae. Digestive gland branches penetrate into the cerata. Cerata flattened, translucent with opaque white spots, with a characteristic fold in the middle and numerous white glands on the edges.

Geographic Range: Florida, Costa Rica, Curaçao, Bonaire, Jamaica, Virgin Islands, Guadeloupe, Barbados.

Natural History: This is a nocturnal species. It autotomizes the cerata very easily when disturbed.

Caliphylla mediterranea A. Costa 1867b: 137.

Description: Background color translucent with dark green dense branches of the digestive system entering the head, rhinophores, and cerata, giving the animal a dark green appearance. Cerata translucent, elongate, with highly ramified digestive branches.

Geographic Range: Amphiatlantic; Western Atlantic: Florida, Curaçao, Virgin Islands, Trinidad & Tobago, Brazil.

Natural History: Feeds on the filamentous green alga *Bryopsis plumosa*.

Caliphylla sp.

Size: Up to 20 mm long

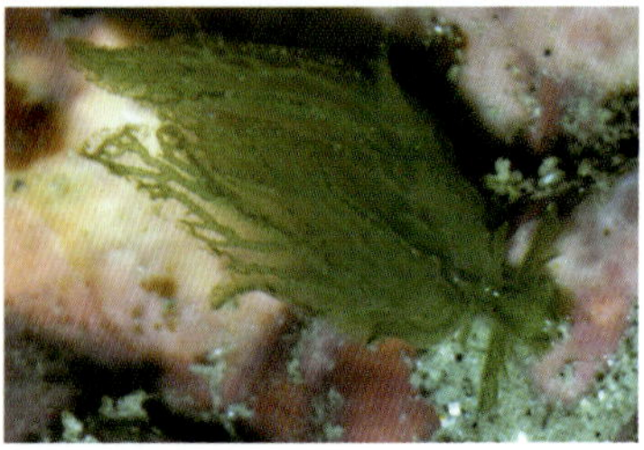

photos J. Hamann & A. Valdés - left: Tobago (JH); right: Costa Rica (AV)

Mourgona murca
Er. Marcus & Ev. Marcus, 1970

Size: Up to 11 mm long

photo J. Hamann - Lameshur Bay, Virgin Islands

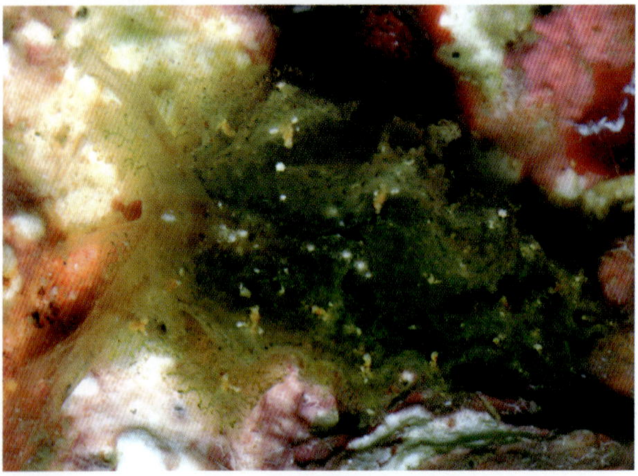

Mourgona germaineae
Er. Marcus & Ev. Marcus, 1970

Size: Up to 15 mm long

photo A. Valdés - Puerto Vargas, Costa Rica

Description: Background color translucent with dense light green branches of the digestive system entering the head, rhinophores, and cerata, giving the animal a green appearance. Cerata translucent, very elongate, with highly ramified digestive branches.

Geographic Range: Costa Rica, Tobago.

Natural History: Feeds on filamentous green algae.

Remarks: This species differs from *Caliphylla mediterranea* by having more elongate cerata and a lighter color.

Mourgona murca Er. Marcus & Ev. Marcus 1970: 34, figs. 53-62.

Description: Background color translucent greyish covered with abundant opaque white pigment. Viscera showing through as a brown area. Cerata translucent, leaf-shaped, with round edges, green branches of the digestive gland and opaque white and brown spots. Rhinophores very long in relation to the body, translucent, covered with brown spots.

Geographic Range: Curaçao, Virgin Islands.

Mourgona germaineae Er. Marcus & Ev. Marcus 1970: 34-35, figs. 63-68.

Description: Background color translucent, with a few opaque white spots. Viscera showing through as a light brown area. Cerata translucent, leaf-shaped with irregular edges, having numerous glands. Green branches of the digestive gland visible through, giving the cerata a greenish appearance. Translucent greyish rhinophores, very long in relation to the body.

Geographic Range: Florida, Costa Rica, Puerto Rico, Bahamas.

Natural History: Feeds on *Cymopolia barbata*.

Family APLYSIIDAE Lamarck, 1809

Diagnosis: Body oval to elongate, with two parapodia that cover the mantle cavity. In some genera the parapodia are reduced and fused together. Gill situated in the mantle cavity along with the purple gland and the opaline gland. Anterior end of the body with a pair of rhinophores and a pair of oral tentacles. Shell internal or absent. Digestive system with two jaws composed or rodlets and cuticularized gizzard plates. Radula composed of numerous, denticulate lateral teeth and a median tooth in each row with a central cusp and several denticles.

Biology: Aplysiids are generally large animals, some species reaching up to 750 mm in length, and they can weigh well over 2 kg. Most species are cosmopolitan and found in temperate and tropical seas, inhabiting shallow coastal areas and sheltered bays, covered with vegetation. All aplysiids are herbivorous, eating a variety of red, green or brown algae and sea grasses. Their color is diet-derived from the pigments of the algae. They concentrate toxins extracted from the algae. Many species secrete ink when disturbed or attacked; there are two main secretory glands in the mantle cavity: the purple gland, situated on top of the mantle cavity, above the gill, produces a red or purple ink, and the opaline gland, situated on the floor of the mantle cavity, beneath the gill, produces a white opaque secretion. Some species mate in large numbers forming chains of up to 20 sea hares mating together. The one at the front acts as a female and the one at the back as a male; all the animal(s) in between are acting as both males and females.

Remarks: The systematics of the family Aplysiidae in the Caribbean is in need of revisionary work. Descriptions of numerous species are based on preserved specimens and very little is known about the variability in external coloration. Most species of aplysiids have large distribution ranges and include populations that are no longer in physical contact, which may indicate that they actually constitute different species.

photo J. Hamann - Point du Bout, Martinique

Aplysia parvula
Mörch, 1863

"Small seahare"

Size: Up to 30 mm long

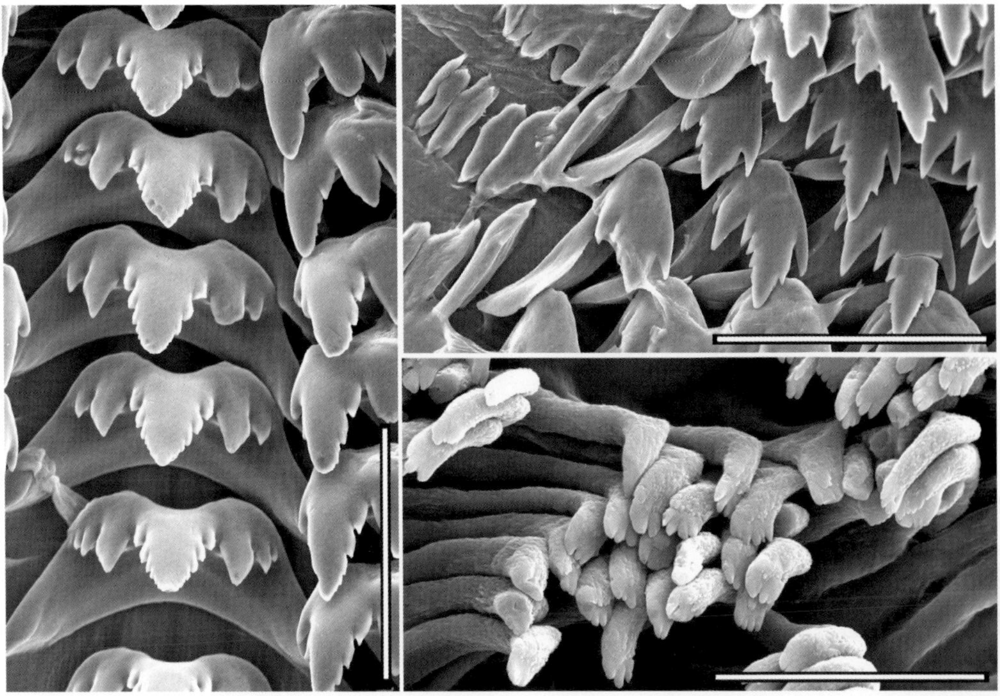

photos A. Valdés - SEM photos of *Aplysia parvula*; left: central radular teeth, scale bar = 50 µm; upper right: outer radular teeth, scale bar = 50 µm; lower right: jaw elements, scale bar = 30 µm

Aplysia parvula Mörch 1863: 22-23.
[numerous synonyms from this and many other geographic areas not listed]

Description: Background color variable, usually brown or green mottled with whitish blotches, but sometimes more uniform. Large parapodial flaps which have a black or navy blue edge. Rolled rhinophores; tentacles edged with black or navy blue.

Geographic Range: Circumtropical; Western Atlantic: Florida, Mexico, Belize, Honduras, Costa Rica, Aruba, Curaçao, Bonaire, Bermuda, Cayman Islands, Jamaica, Puerto Rico, Virgin Islands, St. Lucia, Martinique, Guadeloupe, St. Vincent & the Grenadines, Barbados, Trinidad & Tobago, Brazil.

Natural History: Nocturnal species found in grassy areas and rocky areas roaming around various species of algae.

Remarks: This is the smallest species of *Aplysia*.

Aplysia juliana
Quoy & Gaimard, 1832

"Walking seahare"

Size: Up to 150 mm long

photo J. Hamann - La Vielle Point, St. Lucia

Aplysia dactylomela
Rang, 1828

"Spotted seahare"

Size: Up to 200 mm long

photo D. Behrens - Roatán, Honduras

Aplysia brasiliana
Rang, 1828

Size: Up to 270 mm long

photos J. Hamann - Venezuela; left: light specimen crawling; right: dark specimen swimming

Aplysia juliana Quoy & Gaimard 1832-33[1832]: 309-310, pl. 24, figs. 5-6.
[numerous synonyms from this and many other geographic areas not listed]

Description: Background color variable, normally brownish with numerous pale spots. Posterior end of the foot forming a temporary sucker, so when the animal crawls along it can use the sucker to anchor itself then greatly extend its head and neck like a leech.

Geographic Range: Circumtropical; Western Atlantic: Florida, Curaçao, St. Lucia, Barbados, Brazil.

Natural History: This is one of the few species of *Aplysia* lacking the ink gland that produces reddish purple ink; instead it produces a milky white secretion.

Aplysia dactylomela Rang 1828: 56, pl. 9.
[numerous synonyms from this and many other geographic areas not listed]

Description: Background color usually greenish brown with large dark or black rings, and blackish reticulate lines. Black markings found both inside and outside of the parapodia. Some specimens with relatively thin black lines and rings. The body is relatively tough with leathery parapodia.

Geographic Range: Circumtropical; Western Atlantic: Florida, Texas, Mexico, Belize, Honduras, Costa Rica, Panama, Colombia, Venezuela, Aruba, Curaçao, Bonaire, Bermuda, Cayman Islands, Jamaica, Puerto Rico, Virgin Islands, Barbados, St. Maarten/St. Martin, St. Lucia, Guadeloupe, Martinique, St. Vincent & the Grenadines, Grenada, Trinidad & Tobago, Brazil.

Natural History: This species produces purple ink when disturbed. Animals mate in chains.

Aplysia brasiliana Rang 1828: 55-56, pl. 8, figs. 1-3.
[several synonyms from this geographic area not listed]

Description: Background color variable, translucent yellow, brown, gray, green or black, usually with irregular patches and mottling of different shades of brown and green, and often with lighter spots of white or yellow.

Geographic Range: Temperate and tropical Atlantic Ocean; Western Atlantic: New Jersey, Florida, Texas, Costa Rica, Colombia, Venezuela, Bermuda, Aruba, Brazil.

Natural History: Active swimmer by flapping the parapodia. Animals mate in chains.

Remarks: This species is likely a synonym of *Aplysia fasciata* from the eastern Atlantic. *Aplysia willcoxi* and *A. cervina* are probably synonyms as well. A review of the western Atlantic species of *Aplysia* is necessary to determine the status of all these species.

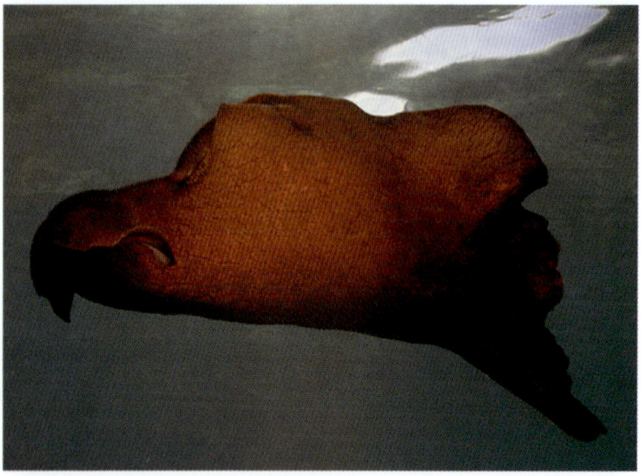

photo A. DuPont - Bahamas

Aplysia morio
(Verrill, 1901)

"Atlantic black seahare"

Size: Up to 400 mm long

photo © L. Ianniello - Dominica

Siphonota geographica
(Adams & Reeve, 1850)

Size: Up to 170 mm long

photo J. Hamann - Dinner Key, Florida

Bursatella leachii
de Blainville, 1817

"Ragged seahare"

Size: Up to 120 mm long

Tethys morio Verrill 1901: 25-26, pl. 3, figs. 5-5A.
Aplysia donca Ev. Marcus & Er. Marcus 1960: 251-253, figs. 1-5.

Description: Background color deep brown to black with no spots. Sometimes darker lines and stripes on head and sides.

Geographic Range: Rhode Island, Georgia, Florida, Texas, Bermuda, Bahamas.

Natural History: Active swimmer by flapping the parapodia.

Remarks: A review of the western Atlantic species of *Aplysia* is necessary to determine the status of this species.

Siphonotus geographicus Adams & Reeve 1850: 64-65, pl. 18, fig. 1.

Description: Background color variable from greenish to pale brown, with a map-like pattern of white pigment. Rhinophores arranged close together and set back between the parapodial lobes.

Geographic Range: Indo-Pacific and Western Atlantic: Florida, Dry Tortugas, St. Lucia, Dominica.

Natural History: Swims by flapping the parapodia. Animals mate in chains.

Remarks: The Latin name *geographicus* refers to the map-like nature of the color pattern.

Bursatella leachii de Blainville 1817: 138.
[numerous synonyms from this and many other geographic areas not listed]

Description: Background color blackish to brown, tan, white or pale lavender, often with white to gold spots. Numerous branched papillae cover the entire body. Body soft.

Geographic Range: Circumtropical; Western Atlantic: North Carolina, Florida, Texas, Belize, Costa Rica, Panama, Colombia, Venezuela, Virgin Islands, Jamaica, Aruba, Curaçao, Bermuda, Trinidad, Brazil.

Natural History: Inhabits sea grass beds and feeds on algae. Dense aggregations have been reported.

Remarks: Several subspecies have been described for this species. Caribbean populations have been assigned to the subspecies *Bursatella leachii pleii* Rang, 1828.

Stylocheilus striatus
(Quoy & Gaimard, 1832)

Size: Up to 45 mm long

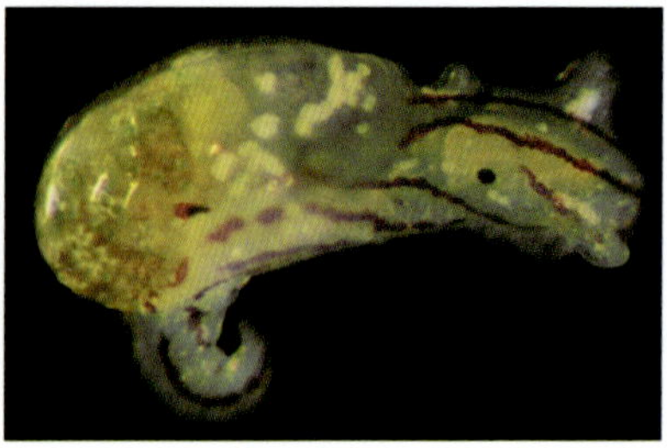

photos J. Hamann & © L. Harris - left: adult, St. Vincent (JH); right: juvenile, Virgin Islands (LH)

Stylocheilus longicaudus
(Quoy & Gaimard, 1824)

Size: Up to 60 mm long

photo © B. Putnam Hansen - Florida

Notarchus punctatus
Philippi, 1836

Size: Up to 32 mm long

photo J. Hamann - Wallilabu, St. Vincent

Aplysia striata Quoy & Gaimard 1832-33[1832]: 315-316, pl. 24, figs. 9-11.
Notarchus polyomma Mörch 1863: 25.

Description: Background color translucent brown to gray, with longitudinal darker lines and scattered spots, either with bright pink or blue centers. Body elongate with numerous branched papillae. Juveniles with an external shell.

Geographic Range: Cosmopolitan; Western Atlantic: Florida, Dry Tortugas, Belize, Colombia, Venezuela, Aruba, Curaçao, Bonaire, Bermuda, Bahamas, Cayman Islands, Jamaica, Puerto Rico, Virgin Islands, Martinique, Barbados, St. Vincent & the Grenadines, Grenada, Brazil.

Natural History: Feeds on algae; common in shallow waters.

Remarks: Records of this species have often been assigned to *Stylocheilus longicauda*, which is a pelagic species found in association with floating algae.

Aplysia longicauda Quoy & Gaimard 1824: 421, 422, pl. 66, fig. 8.
[several synonyms from this and other geographic areas not listed]

Description: Background color bright yellow to bright green with scattered spots. Spots usually blue, often with a reddish edge. White and blue spots without a surrounding ring also occur. Scattered short papillae over the body. Body very elongate with an extremely prolonged posterior end.

Geographic Range: Cosmopolitan; Western Atlantic: Florida, Brazil.

Natural History: Found floating in the open ocean on drifting brown algae. The sole produces a very sticky mucus.

Notarchus punctatus Philippi 1836: 255-256, pl. 7, fig. 9.

Description: Background body translucent gray, with opaque white and occasionally black markings. Viscera visible through the skin. Body globose; parapodia almost completely fused, except for a small anterior opening to allow an exchange of water to the mantle cavity. Shell absent.

Geographic Range: Mediterranean, Japan, and western Atlantic: Florida, Belize, Costa Rica, Jamaica, Guadeloupe, St. Vincent & the Grenadines.

Natural History: This species is able to escape from danger swimming by jet propulsion, tumbling through the water like an out of control party balloon.

Dolabrifera dolabrifera
(Cuvier, 1817)

"Warty seahare"

Size: Up to 90 mm long

photo J. Hamann - Grenada

photo J. Hamann - Pidgeon Island, St. Lucia - ventral view

photo A. Valdés - Curaçao

Petalifera petalifera
(Rang, 1828)

"Rang's seahare"

Size: Up to 40 mm long

photo J. Hamann - Fort St. Louise, Martinique

Dolabella dolabrifera Cuvier 1817: 416.
[numerous synonyms from this and many other geographic areas not listed]

Description: Background color varies from a mottled green to light brown or even pink to dark brown. Body covered by low tubercles. Animals flattened, tapered in shape, the posterior half usually broader and rounded, gradually narrowing towards the head. Parapodia fused except for a short region in the posterior midline.

Geographic Range: Circumtropical; Western Atlantic: Florida, Costa Rica, Colombia, Aruba, Curaçao, Bonaire, Venezuela, Bermuda, Bahamas, Cayman Islands, Puerto Rico, Jamaica, Virgin Islands, Saba, Antigua, Martinique, St. Lucia, Grenada, Barbados, St. Vincent & the Grenadines, Brazil.

Natural History: Feeds on small algal coverage on rocks. Very abundant locally on intertidal rocky areas. It crawls with a leach-like movement.

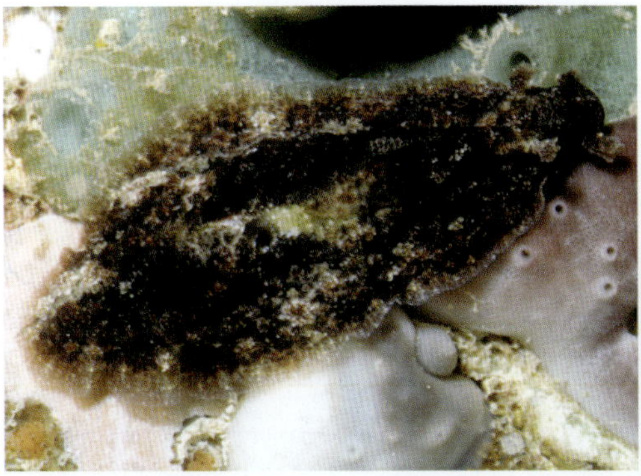

photos J. Hamann - Pidgeon Island, St. Lucia

Aplysia petalifera Rang 1828: 52, pl. 5, figs. 1-3.
[numerous synonyms from this and many other geographic areas not listed]

Description: Background color translucent with green to brown specks or lines and opaque white pigment. The translucent body enables this species to be well-camouflaged on the sea grass leaves and algae on which it lives. The color of the plant shows through the body of the slug. Parapodia reduced and fused, except for a small posterio-dorsal opening into the mantle cavity. Dorsum with ramified papillae.

Geographic Range: Circumtropical; Western Atlantic: Puerto Rico, Bonaire, Barbados, Martinique, Brazil.

Natural History: Feeds on sea grasses and algae.

photo J. Hamann - Cayo Luis Peña, Puerto Rico

Petalifera ramosa
Baba, 1959

"Baba's petalifera"

Size: Up to 70 mm long

photo J. Hamann - Isla Mujeres, Mexico

Phyllaplysia engeli
Er. Marcus, 1955

"Engel's seaslug"

Size: Up to 15 mm long

photo J. Hamann - Grand Bahama

Phyllaplysia smaragda
Clark, 1977

Size: Up to 22 mm long

Petalifera ramosa Baba 1959: 43-44, figs. 1-5.
Petalifera habei Eales 1960: 12-18, figs. 1-6.

Description: Background color translucent greenish, with opaque white pigment on different areas of the dorsum. Firm, conical tubercles scattered over the body; smaller ones often pink with a darker basal ring, larger ones pale brownish white with a dark brown basal ring. Some of the larger tubercles have a terminal papilla that often has long filamentous branches.

Geographic Range: Circumtropical, Western Atlantic: Florida, Mexico, Belize, Honduras, Costa Rica, Colombia, Cuba, Puerto Rico, Jamaica, Martinique, Trinidad & Tobago.

Natural History: When disturbed this species can swim by dorso-ventral flexion of the body.

Phyllaplysia engeli Er. Marcus 1955: 105-108, figs. 29-37.

Description: Background color translucent with spotting and patching of pink, brown, white and shadows of green, the proportions varying between animals. Some specimens have white longitudinal lines, and some have low papillae. Body flattened, very cryptic on the leaves of sea grasses on which it is found.

Geographic Range: Florida, Mexico, Costa Rica, Colombia, Curaçao, Bahamas, Puerto Rico, Jamaica, St. Maarten/St. Martin, Barbados, Brazil.

Natural History: Found on the sea grasses *Thallasia testudinum* and *Halodule* spp.

Phyllaplysia smaragda Clark 1977: 651-657, figs. 2-4.

Description: Background color grass to emerald green with gray-pink patches on the left side of dorsum. Rhinophores and oral tentacles with greyish pigment. Dorsum with minute tubercles.

Geographic Range: Florida, Bahamas.

Natural History: Found on sea grasses in shallow water.

Remarks: This species is distinguishable from *Phyllaplysia engeli* by having a much more narrow and elongate body and numerous anatomical differences discussed in Clark (1977).

Families UMBRACULIDAE Dall, 1889 & TYLODINIDAE Gray, 1827

Diagnosis: Body oval, with an external, limpet-like shell covering the body completely (tylodinids) or partially (umbraculids). Shell oval to rounded, with a subcentral, conical apex bearing the larval shell (protoconch). Body covered with tubercles in umbraculids and smooth in tylodinids. Both groups with two oral tentacles and two rhinophores and the gill on the left side of the body. Umbraculids with a deep mid-anterior cleft, tylodinids with an external sperm groove that leads from the genital aperture at the base of the right rhinophore. Penis retractile in tylodinids and non retractable in umbraculids. Anus posterior in umbraculids and lateral in tylodinids. Oral tube with numerous, irregular, thickened polygons. Radula broad, with numerous lateral teeth (up to 800,000 in *Umbraculum*) and a median tooth (tylodinids). Muscular part of the stomach with cuticularized longitudinal ridges.

photo J. Hamann - Bat Cave Point, Martinique

Umbraculum umbraculum
(Lightfoot, 1786)

"Umbrella slug"

Size: Up to 280 mm long

photo © L. Harris - Florida

Tylodina americana
Dall, 1890

Size: Up to 10 mm long

Biology: Members of *Umbraculum* feed upon a diverse range of demospongiid sponges; they are slow but mobile organisms that actively search for food sources. By contrast, tylodinids feed only on sponges of the family Aplysinellidae; while feeding, they make holes in the sponge surface similar in shape and size to a natural siphons. In addition, tylodinids are usually buried in these holes, as they display sedentary behavior over long periods of time. The shell of *Umbraculum* is usually covered with encrusting plant and animal growths.

Remarks: Many authors consider that there is a single species in the family Umbraculidae, *Umbraculum umbraculum*, with a cosmopolitan distribution throughout the tropical and warm temperate Indo-Pacific, Eastern Pacific, and Atlantic, but several species names have been described from different geographic regions. Tylodinidae includes two genera, *Tylodina* and *Anidolyta*, with only one species present in the western Atlantic.

Patella umbracula Lightfoot 1786: 178.
[numerous synonyms from this and many other geographic areas not listed]

Description: Background color variable, black, brown, yellow or orange. Body oval, covered with large dorsal tubercles and a flat, external shell with a central conical elevation that is culminated by the protoconch. Rhinophores smooth extending from beneath the shell.

Geographic Range: Circumtropical; Western Atlantic: Florida, Louisiana, Texas, Colombia, Martinique.

Natural History: This species feeds on sponges and is often found burrowing or under rocks Shell usually covered with encrusting plant and animal growth.

Remarks: Named for the umbrella-like shell.

Tylodina americana Dall 1890: 121-122.

Description: Background color uniformly yellowish. Body oval, covered with large dorsal tubercles and a conical, external shell with a central conical elevation that is culminated by the protoconch. Shell with a reddish periostracum. Rhinophores smooth extending from beneath the shell.

Geographic Range: Florida, Texas, Bermuda, Cuba.

Natural History: Feeds on sponges.

Remarks: This is a very rare species rarely observed alive. The original description and most subsequent records are based on shell material.

Family PLEUROBRANCHIDAE Gray, 1827

Diagnosis: Body oval to elongate. Gill situated on the posterior right side of the body. External shell absent, with or without internal shell. When present internal shell is plate-like. Veil trapezoidal, between the oral tentacles and above the mouth. Rolled oral tentacles and rhinophores, a posterior pedal gland on the foot sole, a median buccal gland, and a retractile penis immediately in front of the gill on the right side. Digestive system with median buccal gland, two jaws composed of interlocking cuticularized elements, and a radula composed of numerous lateral hooked-shaped teeth. Anus opens above the gill.

photo J. Hamann - lateral gill, Venezuela

Biology: Many pleurobranchids produce acidic fluids for their own defense. *Berthella* species

photo J. Hamann - Aruba

Berthellina quadridens
(Mörch, 1863)

Size: Up to 80 mm long

photo J. Hamann - Tortola, Virgin Islands

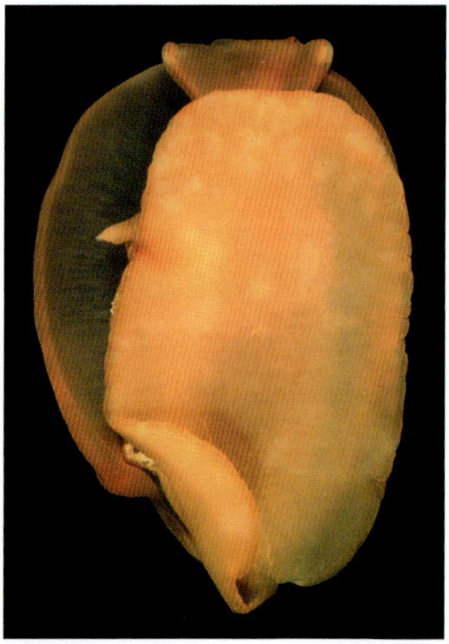

photo © L. Harris - ventral view showing the penis, Guana Island, Virgin Islands feed on sponges, *Pleurobranchus* species feed on ascidians, and other groups have less specialized diets.

Remarks: The systematics of the Caribbean Pleurobranchidae is in need of review. Several species are only known from preserved specimens. Particularly problematic are the genera *Pleurobranchus* and *Pleurobranchea*. An attempt to clarify species of *Pleurobranchus* is provided in the next pages, but this needs to be substantiated with anatomical data. Deep-water taxa have not been included in this book.

photos A. Valdés - SEM photos of *Berthellina quadridens*; left: radular teeth, scale bar = 30 µm; right: jaw elements, scale bar = 50 µm

Berthella quadridens Mörch 1863: 29-31.
Berthella circularis Mörch 1863: 31.
Pleurobranchus amarillius Mattox 1953: 109-114, figs. 1-10.

Description: Background color variable, from translucent yellow, to orange or bright red. Body oval, smooth.

Geographic Range: Mexico, Belize, Colombia, Panama, Venezuela, Aruba, Curaçao, Haiti, Jamaica, Puerto Rico, Virgin Islands, St. Maarten/St. Martin, St. Lucia, Guadeloupe, Martinique, Barbados, St. Vincent & the Grenadines, Grenada, Trinidad & Tobago, Brazil.

Remarks: Thompson (1977) synonymized *Berthella circularis* with *B. quadridens*, whereas Ev. Marcus (1984) considered them to be different species based on radular differences. Those differences are minor and not correlated with external morphological variation, therefore we follow Thompson's opinion. Similarly colored species from other geographic regions belong to different species (Gosliner & Cervera, personal communication).

photos J. Hamann & A. Valdés - left: head, Virgin Islands (JH); right: juvenile, Curaçao (AV)

Berthella stellata
(Risso, 1826)

Size: Up to 20 mm long

photos J. Hamann - Bahamas; left: adult specimen; right: specimen laying eggs.

Berthella agassizii
(MacFarland, 1909)

"Agassiz's berthella"

Size: Up to 25 mm long

photo J. Hamann - Lark Cay, Belize

Berthella sp.

Size: Up to 24 mm long

photo J. Hamann - Puerto Rico

Pleurobranchus stellatus Risso 1826: 41.
[other synonyms from several geographic areas not listed]

Description: Background color translucent white to honey-brown. Dorsum covered with opaque white markings; an opaque white transverse bar or cross is generally situated near the middle of the notum, but may be absent in some specimens.

Geographic Range: Circumtropical; Western Atlantic: Florida, Belize, Honduras, Panama, Colombia, Venezuela, Bahamas, Curaçao, Cayman Islands, Puerto Rico, Virgin Islands, St. Lucia, Martinique, Grenada, Brazil.

Natural History: Found intertidally or in the shallow subtidal zone under rocks.

Remarks: Named for star-like marking on dorsum. *Berthella tupala* Er. Marcus, 1957 from Brazil is a synonym.

Pleurobranchus agassizii MacFarland 1909: 59-64, pls. 11-12, figs. 43-57.

Description: Background color pink to reddish brown, with scattered opaque white spots. Body oval.

Geographic Range: Belize, Colombia, Curaçao, Bermuda, Brazil.

photo J. Hamann - Lark Cay, Belize

Description: Background color yellowish with numerous minute brown dots and a series of white patches. Dorsum covered with opaque white markings; a black spot is situated on the center of the notum.

Geographic Range: Puerto Rico

Remarks: This is probably an undescribed species.

Pleurobranchea inconspicua
Bergh, 1897

Size: Up to 60 mm long

photo J. Hamann - Cooper Island, Virgin Islands

Pleurobranchus areolatus
Mörch, 1863

"Atlantic sidegill slug"

Size: Up to 150 mm long

photos J. Hamann - left: adult specimen, Aruba; right: juvenile specimen, St. Maarten/St. Martin

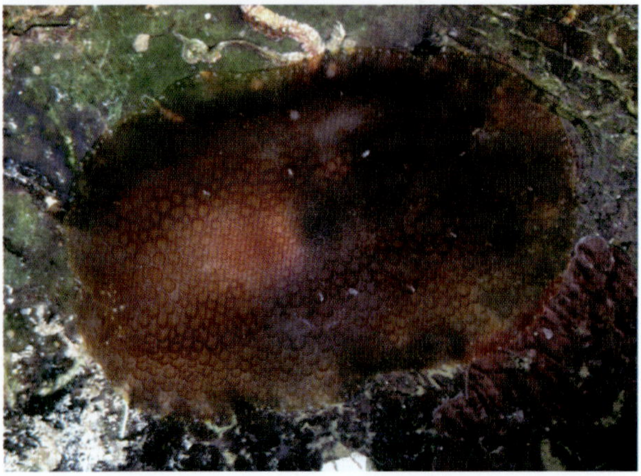

Pleurobranchus evelinae
Thompson, 1977

"Eveline's sea slug"

Size: Up to 60 mm long

photo A. Valdés - Bocas del Toro, Panama

Pleurobranchea inconspicua Bergh 1897: 49-51, pl. 8, figs. 2-10.
Pleurobranchea hedgpethi Abbott 1952: 1-2, pl. 1, figs. 1-8.
Pleurobranchea hamva Ev. Marcus & Er. Marcus 1957: 21-25, figs. 40-52.
Pleurobranchea gela Ev. Marcus & Er. Marcus 1966: 174-177, figs. 35-37.

Description: Background translucent white with a brown reticulate pattern and large opaque white spots. Rhinophores far apart and veil with small papillae. Mantle with a siphon at the back and posterior end of the foot with a spur.

Geographic Range: West Africa and Western Atlantic: Florida, Belize, Honduras, Venezuela, Virgin Islands, Martinique, St. Vincent & the Grenadines, Grenada, Brazil.

Natural History: Found on sandy bottoms from 0-100 m depth.

Remarks: *Pleurobranchea bonnieae* Ev. Marcus & Gosliner, 1984 is likely a synonym.

Pleurobranchus areolatus Mörch 1863: 28-29.
Susania gardineri White 1952: 106-107, pl. 6, fig. 1.

Description: Background color cream with deep reddish brown pigment on the larger tubercles, mantle margin and rhinophores and yellow spots. Tubercles few, wide, rounded, flat, with larger ones surrounded by smaller ones. Rhinophores large, fused together at the base.

Geographic Range: Venezuela, Jamaica, St. Thomas, Aruba, St. Maarten/St. Martin, Brazil.

Remarks: This species is characterized by having large rounded tubercles surrounded by smaller ones and the dark, large rhinophores fused at the base. Because this species has often been misidentified, a restricted geographic range to verified localities is provided.

Pleurobranchus evelinae Thompson 1977: 108-109, figs. 12E-F, 13C-E.

Description: Background color uniformly reddish brown with some white spots. Tubercles numerous, small, polygonal, flat, with the smaller ones on the center of the dorsum. Rhinophores small proportional to the body size, fused together at the base.

Geographic Range: Jamaica, Panama.

Natural History: Probably feeds on ascidians.

Remarks: This species is characterized by having numerous polygonal, flat tubercles, smaller on the center of the dorsum and small rhinophores fused at the base.

Pleurobranchus crossei
Vayssière, 1897

Size: Up to 120 mm long

photo J. Hamann - Boquerón, Puerto Rico

photos J. Hamann & A. DuPont - left: Culebra, Puerto Rico (JH); right: Bahamas (AD)

***Pleurobranchus* sp.**

Size: Up to 100 mm long

photos A. Valdés - Manzanillo, Costa Rica; left: adult specimen; right: detail of the rhinophores

Pleurobranchus crossei Vayssière 1897: 353-354, fig. 1.
Pleurobranchopsis aurantiaca Verrill 1900: 547, pl. 66, fig. 5 [non *Pleurobranchus aurantiacus* Risso, 1818].
Pleurobranchus verrilli Thiele 1931: 419.
Pleurobranchus atlanticus Abbott 1949: 73-77, pl. 5, figs. 1-10D.

Description: Background color white to yellow with deep reddish brown and opaque white pigment on some tubercles; darker tubercles generally have a darker line around them. Color arrangement appears random and not related to the size or position of the tubercles. Tubercles numerous, small, polygonal, conical, all of similar size. Rhinophores large, not fused together at the base, with similar color to the rest of the body. Foot with numerous brown spots.

Geographic Range: Mexico, Costa Rica, Bahamas, Puerto Rico, Bermuda

Remarks: This is the most common *Pleurobranchus* species in the Caribbean. It is distinguishable from *P. evelinae* by having conical tubercles all of similar size. Because this species has often been misidentified, a restricted geographic range to verified localities is provided.

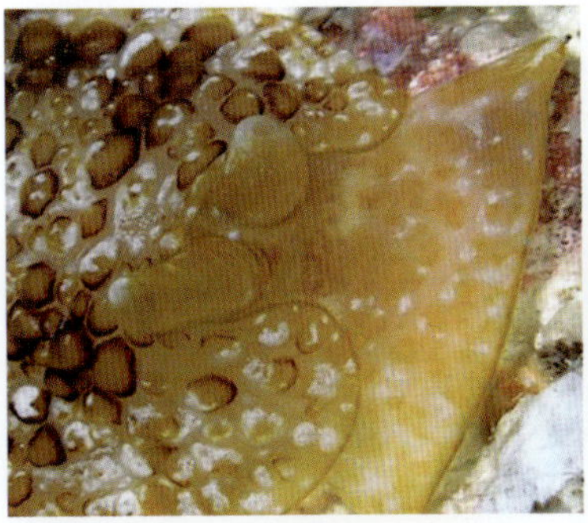

photos A. DuPont - Bahamas; left: rhinophores; right: ventral view of the foot

---

Description: Background color pale gray with irregular dark brown and opaque white pigment. Dorsum covered by numerous white dots. Tubercles large, irregular, clustered together, larger on the center of the dorsum. Rhinophores small proportionally to the size of the animal.

Geographic Range: Florida, Costa Rica.

Remarks: This species, characterized by having large, irregular, clustered tubercles, is probably undescribed.

Family HEXABRANCHIDAE Bergh, 1891

Diagnosis: Dorsum smooth, lacking spicules. Head with two large, flattened and lobate oral tentacles. Anterior border of the foot simple. Gill contractile, not retractile. Radula composed of simple, hamate teeth. Labial cuticle completely covered with rodlets. Buccal mass with numerous and strong muscles attached. Penis and vagina devoid of hooks. Vestibular or accessory glands absent.

Biology: Members of this family feed on sponges and are exclusively found in the tropical Indo-Pacific and Caribbean. When disturbed, these animals display a flashing behavior by suddenly exposing brightly colored areas of the dorsum, which are normally covered by folds of the mantle margin; this can be accompanied by a vigorous swimming escaping behavior (see next page). Members of the Hexabranchidae are among the largest nudibranchs reaching up to 600 mm long.

Hexabranchus morsomus
Ev. Marcus & Er. Marcus, 1962

"Caribbean spanish dancer"

Size: Up to 100 mm long

photo J. Hamann - Dog Island, Anguilla

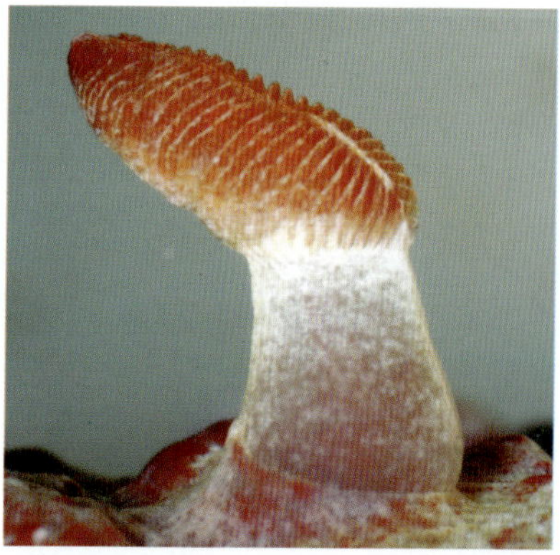

photo A. Valdés - rhinophore, Puerto Viejo, Costa Rica

photo J. Hamann - Guanaja, Puerto Rico

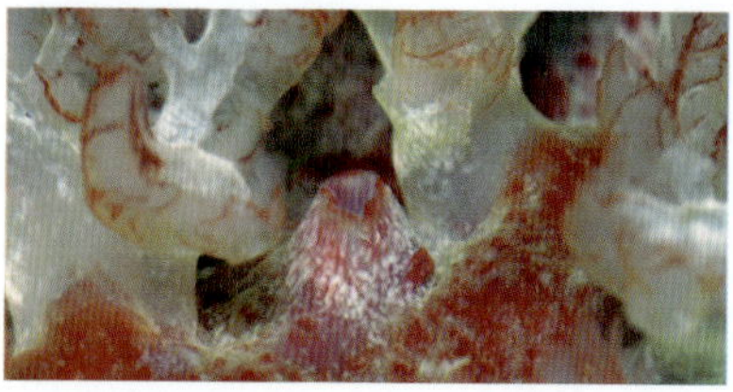

photo A. Valdés - anus, Puerto Viejo, Costa Rica

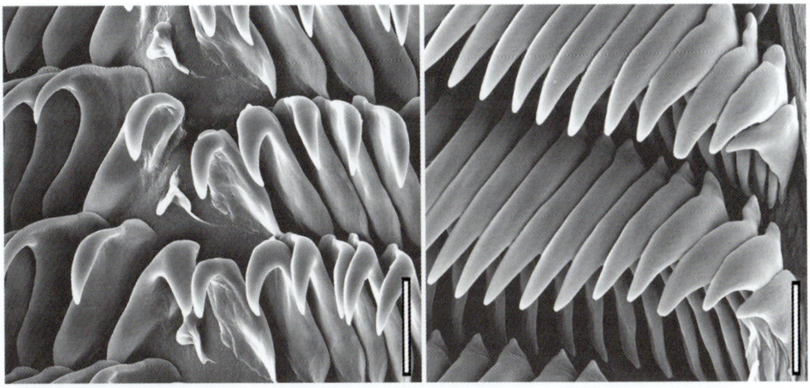

photos A. Valdés - SEM photos of *Hexabranchus morsomus*; left: central radular teeth, scale bar = 60 µm; right: outer radular teeth, scale bar = 60 µm

Hexabranchus morsomus Ev. Marcus and Er. Marcus 1962: 468-471, figs 16-17.

Description: Background color from orange to red with mottled patches and spots of red, orange or yellow. Juvenile specimens whitish. Some specimens can have varying amounts of white, especially on the mantle margin. Rhinophoral clubs can be entirely red or translucent yellow with a red tip. Branchial leaves usually red, tipped with white, but in smaller specimens can be translucent yellow with red tips.

Geographic Range: Honduras, Panama, Costa Rica, Venezuela, Aruba, Puerto Rico, Virgin Islands, St. Maarten/St. Martin, St. Lucia, Martinique, Antigua, Grenada, St. Vincent & the Grenadines, Trinidad & Tobago.

Natural History: Inhabits living reefs down to 33 m depth.

Remarks: The new genus name *Caribranchus* has been recently introduced for *Hexabranchus morsomus* (Ortea *et al*. 2002), but there is little phylogenetic justification for such a change. Therefore *Caribranchus* is here regarded as a synonym of *Hexabranchus*.

photos J. Hamann - left: gill, Dog Island, Anguilla; right: juvenile specimen, St. Maarten/St. Martin

Swimming behavior

Species of *Hexabranchus* display a complex swimming behavior, which consists of a combination of a dorso-ventral flexing of the entire body and a muscular wave that runs along each side of the mantle, causing the wing-like flapping of the mantle edge. The common name of the species derives from these dancing-like movements that resemble the performance of a female flamenco dancer from southern Spain.

photo A. Valdés - Puerto Viejo, Costa Rica

Family AEGIRETIDAE Fischer, 1883

Diagnosis: Body elongate, heavily spiculose, with a reduced, indistinct mantle margin. Dorsum covered with numerous tubercles variable in shape from simple, rounded bumps to elaborate pedunculate papillae. Rhinophores emerging from a cylindrical rhinophoral sheath with tubercles around the edge. Gill protected by tuberculate lobes. Jaws consist of a cuticularized plate devoid of rodlets. Radula with numerous simple, hamate lateral teeth; median teeth absent.

Biology: Species of the Aegiretidae feed exclusively on calcareous sponges. Animals vary considerably in size, from 15 mm to over 140 mm in some tropical Indo-Pacific species.

Aegires gomezi

Ortea, Luque, & Templado, 1990

Size: Up to 4 mm long

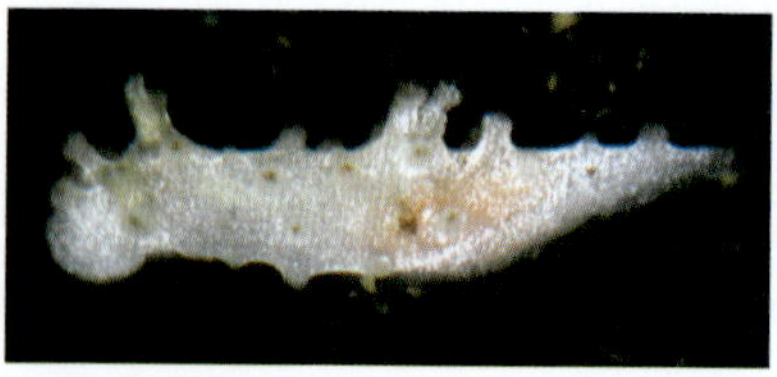

photos J. Hamann & A. Valdés - left: Cayman Islands (JH); right: Costa Rica (AV)

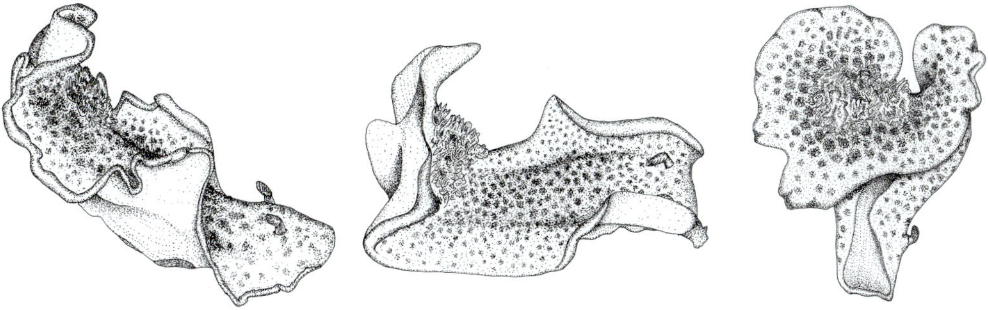

drawings D. Behrens - Series of positions displayed by species of *Hexabranchus morsomus* during swimming.

photos A. Valdés - SEM of *Aegires sublaevis*; left: inner radular teeth, scale bar = 50 µm; center: mid-lateral radular teeth, scale bar = 50 µm; right: outer radular teeth, scale bar = 50 µm

Aegires gomezi Ortea, Luque, & Templado 1990: 333-336, figs. 1B-C, 2A-B.

Description: Background color translucent white with scattered opaque white spots all over the dorsum. Dorsal papillae stalked with a flattened mushroom like tip; they are arranged in two lateral rows, and another two rows between the rhinophores and the gill. Behind the gill there is a single median row. Flattened tip of the tubercles translucent with a central dark brown spot.

Geographic Range: Florida, Costa Rica, Cayman Islands, Cuba.

photo J. Hamann - Grand Bahama, juvenile

photo J. Hamann - Miami, Florida

Aegires ortizi
Templado, Luque, & Ortea, 1987

Size: Up to 8 mm long

photo J. Hamann - Mosquito Bank, Florida

Aegires sublaevis
Odhner, 1932

Size: Up to 12 mm long

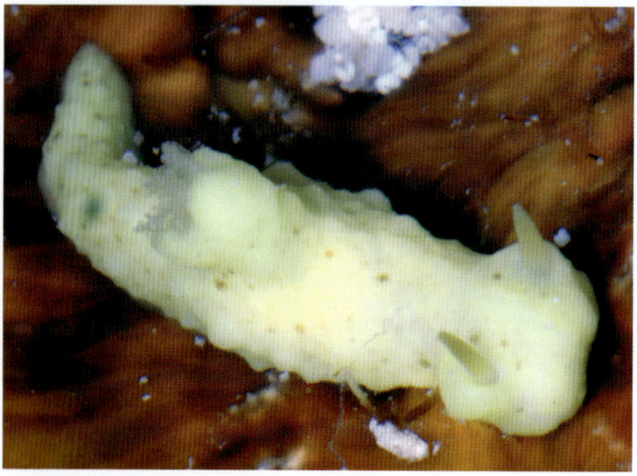

photo A. Valdés - St. Thomas, Virgin Islands

photo J. Hamann - juvenile specimen, Guanaja, Honduras

Aegires ortizi Templado, Luque, & Ortea 1987: 306-307, figs. 1A, 2A, 3A.

Description: Background color variable, some specimens with noticeable brown spots and others with none. Tubercles vary from conical to mushroom shaped, with a flattened top in some individuals suggesting that tubercles can change shape as described for other species of *Aegires*. In paler animals spicules are visible projecting from the tip of some tubercles.

Geographic Range: Cayman Islands, Bahamas, Cuba.

Natural History: Found on algae of the genera *Cladophora* and *Sargassum* in the Bahamas.

Aegires sublaevis Odhner 1932: 39-41, figs. 8-10, pl. 2, figs. 29-31.

Description: Background color variable from lemon yellow to a pale brownish white. Some specimens with large scattered grayish brown spots, each edged with a ring of opaque background coloration. Dorsum with a few stalked dorsal tubercles and some rounded bumps scattered all over.

Geographic Range: Eastern Atlantic and Western Atlantic: Florida, Honduras, Costa Rica, Panama, Virgin Islands, Bermuda.

Natural History: Reported to feed on the yellow sponge *Clathrina coriacea*.

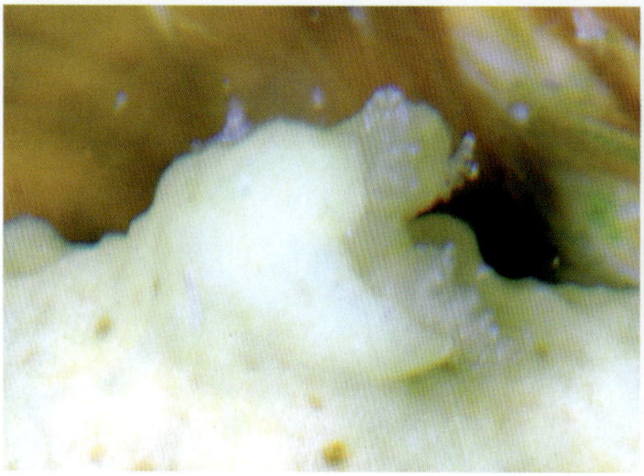

photos A. Valdés - St. Thomas, Virgin Islands; left: gill; right: rhinophores

Families CORAMBIDAE Bergh, 1871 & ONCHIDORIDIDAE Gray, 1827

Diagnoses: Both families with a rounded to oval, flattened body. Rhinophores retractable into individual pockets; smooth or with one or two curved lamellae on each side of the central rod in Corambidae and lamellate in Onchidorididae. Corambidae with several ventral, simple brachial leaves, situated below the mantle margin on each side of the anus. Onchidorididae with posterio-dorsal gill and anus; branchial leaves non-retractile. Some species of Corambidae with the posterior mantle margin raised into a pallial notch. Anterior end of the foot with a ventral velum covered by the mantle margin, normally bilobed or projected laterally into a pair of tentacles. Digestive system with a chitinous cuticle lacking rodlets. A muscular buccal pump opens off the buccal bulb. Radula lacking median teeth, except for some species of Onchidorididae. Innermost lateral tooth on each side large, usually with a triangular base and a recurved cusp that may have denticles. Other lateral teeth are much smaller and non-denticulate.

photo J. Hamann - Grand Bahama

Corambe obscura
Verrill, 1870

"Obscure corambe"

Size: Up to 8 mm long

photo J. Hamann - Halifax, Grenada

Diaphorodoris sp.

Size: Up to 9 mm long

Biology: Corambids feed on species of bryozoans of the genus *Membranipora* and are cryptically colored matching the pattern of their prey; some species have a white reticulate pattern matching the pattern of the bryozoan. Species of Onchidorididae also feed on encrusting bryozoans except for one species that feeds on barnacles.

Remarks: Members of the Onchidorididae are mostly found in temperate and cold waters. Several species of *Onchidoris* have been described from Bermuda, but their status remain uncertain. The only species found in the Caribbean Sea is the *Diaphorodoris* described below. The genus *Diaphorodoris* includes some tropical species found in the Indo-Pacific and the eastern Atlantic.

Most species of Corambidae probably have circumglobal distributions, and have accompanied the spread of their host bryozoans on floating objects and debris.

Corambe obscura Verrill 1870: 408, figs. 2-3.
[numerous synonyms from this and other geographic areas not listed]

Description: Background color translucent white so it picks up the background color of the surface on which it lives. Internal organs appear yellowish through the dorsum and orange-brown through the foot. Dorsum covered by small dark spots.

Geographic Range: Widely distributed in the Atlantic Ocean, Western Atlantic: from Massachusetts to Brazil.

Natural History: Found under rocks and on floating algae, associated with bryozoans.

Description: Background color milky white with a series of yellowish orange spots on the dorsum. Mantle margin surrounded by a yellow line. Dorsum covered with long conical tubercles. Rhinophores white with an ochre subapical area. Gill white.

Geographic Range: Grenada.

Natural History: Found at 30 m depth.

Remarks: This species is probably undescribed and it is the first record of the genus for the Caribbean.

Family GONIODORIDIDAE H. Adams & A. Adams, 1854

Diagnosis: Body oval to elongate, with the mantle margin reduced to a lateral ridge with or without a few tentacular processes. Rhinophores contractile, not retractile, long, with a few small lamellae. Gill composed of simple branchial leaves, also contractile. Head with a large oral veil. Dorsum with a mid-dorsal ridge or papillae of various sizes. Foot larger than the dorsum. Jaws composed of small elements. A muscular buccal pump, sometimes pedunculate, opens off the dorsal mid-line of the buccal bulb; this organ is used to suck food into the mouth. Radula with or without central teeth, which consists of a reduced plate; innermost lateral tooth of each row with a large, recurved cusp, denticulate along its inner edge; outer lateral teeth often denticulate, but much smaller than the inner lateral teeth.

Biology: Species of Goniodorididae feed on compound ascidian colonies or on encrusting of arborescent bryozoans. Many species show remarkable adaptations in color and shape which camouflage them when on their prey.

photo J. Hamann - Cancún, Mexico

Goniodoris mimula
Er. Marcus, 1955

Size: Up to 10 mm long

photo J. Hamann - Piton Bay, St. Lucia

Okenia impexa
Er. Marcus, 1957

“Disheveled okenia”

Size: Up to 7 mm long

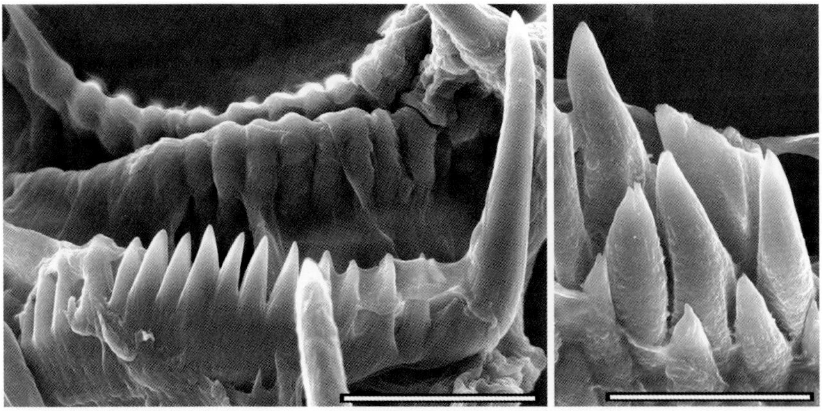

photos A. Valdés - SEM of *Trapania maringa*; left: radular teeth, scale bar = 20 µm; right: jaw elements, scale bar = 20 µm

Goniodoris mimula Er. Marcus 1955: 161-163, figs. 203-208.

Description: Background color pale brown with irregular patches of opaque white and dark brown. Mantle margin narrow, smooth, with no appendages. Velum with two long oral tentacles. Rhinophores and gill cream-white with irregular brown and opaque white patches. Gill composed of numerous gray leaves with white tips.

Geographic Range: Brazil, Mexico.

Natural History: Intertidal to 10 m depth. It feeds on encrusting ascidians.

Remarks: This is the second record of this species originally described from southern Brazil.

Okenia impexa Er. Marcus 1957: 434-438, figs. 120-127.
Cargoa cupella Vogel & Schultz 1970: 390-392, figs. 1-5.
Okenia pusilla Sordi 1974: 1-5, figs. 1-2.

Description: Background color translucent gray, with yellow pigment all over the body. Some specimens with a large black spot between the rhinophores. Rhinophores with a few lamellae. Papillae elongate, pike-shaped, with about five of them along each side of the body, and another located near the center of the dorsum. Gill having the same color as the rest of the body.

Geographic Range: Amphiatlantic; Western Atlantic: Virginia, North Carolina, Puerto Rico, St. Lucia, Brazil.

Natural History: Found attached on *Sargassum* and *Cladophora*.

photo J. Hamann - Chimana Grande, Venezuela

Okenia zoobotryon
(Smallwood, 1910)

"Smallwood's okenia"

Size: Up to 8 mm long

photo J. Hamann - Chimana Grande, Venezuela

Okenia evelinae
Er. Marcus, 1957

Size: Up to 8 mm long

photo J. Hamann - Flamingo Bay, Virgin Islands

Okenia sp. 1

Size: Up to 4 mm long

Polycerella zoobotryon Smallwood 1910: 143-145, fig. 10.
Bermudella polycerelloides Ortea & Bouchet 1983: 50-54, figs. 1-2.

Description: Background color translucent with a series of dark purple and pale brown spots all over the body. A few, larger opaque white spots on the dorsum and laterals of the body. Body elongate with numerous tentacular papillae bearing opaque white spots. Rhinophores lamellate, with the lower half similarly colored to the rest of the body and the upper half covered with opaque white pigment. Gill with the same color as the rest of the body.

Geographic Range: Amphiatlantic; Western Atlantic: Venezuela, Bermuda.

Remarks: The animal here illustrated matches the original descriptions of *Okenia zoobotryon* and *Bermudella polycerelloides*, but not those of subsequent references to *O. zoobotryon*, which should probably be assigned to *Okenia* sp. 2.

Okenia evelinae Er. Marcus 1957: 438-442, figs. 128-137.

Description: Background color translucent white with opaque white spots and purple pigment on the dorsum. Up to four dorsal papillae in front of the gill, two median and two lateral, and a series of up to six blunt, short appendages on each side along the mantle margin. Dorsal appendages, gill, and the borders of the foot are pure white. Rhinophores very long, purple, with lamellae on the posterior side. Branchial leaves short, white.

Geographic Range: Florida, Venezuela, Brazil.

Natural History: Probably feeds on the arborescent bryozoan *Amathia convoluta*.

Description: Background color yellow with opaque white pigment on the dorsum and posterior end of the foot. Up to five wide, tentacular papillae along the mantle edge and a single papilla on the center of the dorsum. Rhinophores yellow, short with lamellae on the posterior side. Gill having the same color as the lateral papillae.

Geographic Range: Bahamas, Virgin Islands.

Remarks: This species appears to be undescribed.

photo J. Hamann - Bahamas

Okenia sp. 2

Size: Up to 6 mm long

photo J. Hamann - Bahamas

Okenia sp. 3

Size: Up to 6 mm long

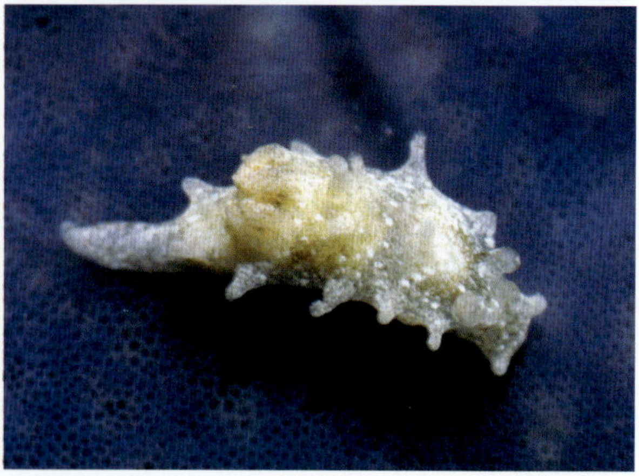

photos J. Hamann - Phosphorescent Bay, Puerto Rico

Description: Background color translucent white with small irregular brown spots scattered over the mantle. Foot and sides of the body may have a few small scattered opaque white spots. Dorsum with up to nine papillae in front of the gill and about 6 on each side along the mantle ridge; papillae with a thin stalk and a rounded terminal apex. Dorsum with a characteristic hump in front of the gill. Rhinophores with cup-like folds on the posterior side. Gill composed of elongate leaves.

Geographic Range: Florida, Mexico, Belize, Bermuda, Bahamas, Jamaica, Puerto Rico, Martinique.

Natural History: Found on the bryozoan *Zoobotryon verticillatum* on which it feeds.

Remarks: This species has been confused with *Okenia zoobotryon* but is clearly distinguishable by having a conspicuous hump in front of the gill and a shorter body.

photos J. Hamann - left: Isla Contoy, Mexico; right: Blue Fields, Belize

---

Description: Background color translucent white with irregular brown pigment and some opaque white spots. Up to four stalked, short papillae along the mantle edge and a four papillae on the center of the dorsum. Rhinophores short with lamellae on the posterior side.

Geographic Range: Puerto Rico.

Remarks: This animal appears to be a different species from both *Okenia zoobotryon* and *Okenia* sp. 2. It lacks the characteristic hump of *Okenia* sp. 2 and the papillae shape and color pattern are different from those of *O. zoobotryon*.

photo J. Hamann - Martinique

Trapania dalva
Er. Marcus, 1972

"Florida trapania"

Size: Up to 12 mm long

photo J. Hamann - Halifax, Grenada

Trapania maringa
Er. Marcus, 1957

Size: Up to 10 mm long

photo J. Hamann - San Blás, Panama

Trapania sp.

Size: Up to 10 mm long

Trapania dalva Er. Marcus 1972: 296-297, 299, figs. 23-26.

Description: Background color dark brown with a symmetrical pattern of creamy white blotches down the dorsal mid-line and a pair of diagonal lines running from the mid-line in front of the gills to the outside of the rhinophoral pocket. Brown region spotted with orange-yellow, white spots also present on the sides of the body. Two lateral papillae on the sides of the rhinophores and two on the sides of the gill. Rhinophores creamy white with brown spots. Gill composed of elogate, white leaves.

Geographic Range: Florida, Bahamas, Martinique, Jamaica.

Trapania maringa Er. Marcus 1957: 442-443, figs. 138-145.

Description: Background color yellowish cream with dense brown pigmentation all over the body and a few opaque white spots. Two lateral papillae on each side of the rhinophores and two on each side of the gill. Oral tentacles, rhinophores, gill, and papillae the same color as the rest of the body.

Geographic Range: Grenada, Brazil.

Description: Background color translucent white with dense opaque white pigmentation all over the body and a few brown white spots. Two lateral papillae on the sides of the rhinophores and two on the sides of the gill. Oral tentacles, rhinophores, gill, and papillae the same color as the rest of the body. Oral tentacles with a brown ring at mid-length.

Geographic Range: Panama.

Remarks: This is probably an undescribed species.

Family POLYCERIDAE Alder & Hancock, 1845

Diagnosis: Body elongate with a reduced mantle margin, usually consisting of a veil around the head and sometimes a ridge along the sides of the body. Veil, dorsum, and mantle margin with or without papillae. Papillae simple or branched. Rhinophores lamellate, retractile into a pocket. Gill contractile. Digestive system with a pair of chitinous jaws often with a wing-like masticatory process. Radula with or without central teeth, some groups with a broad medial plate. Lateral teeth very variable between groups, normally differentiated; second lateral teeth normally the largest, bicuspidate; outer lateral teeth normally simple plates. Penis usually armed with hooks.

photo J. Hamann - light organs in *Plocamopherus lucayensis*

Biology: Members of the Polyceridae feed on encrusting or arborescent bryozoans, colonial ascidians, or oth-

photo © S. Kaicher - Florida

Polycera hummi
Abbott, 1952

Size: Up to 20 mm long

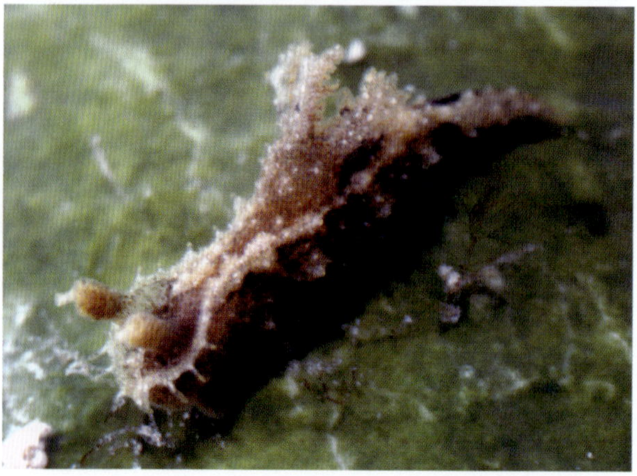

Polycera herthae
Ev. Marcus & Er. Marcus, 1963

"Hertha's polycera"

Size: Up to 9 mm long

photos J. Hamann - Miami, Florida; left adult specimen; right: two juvenile specimens er nudibranchs. Several species have cosmopolitan distributions as a result of their relationship with fouling bryozoan colonies found on ships and floating debris. Species of *Plocamopherus* and *Kaloplocamus* are able to swim by lateral flexions of the body; the posterior end of the foot has a long median crest that functions as a paddle. Species of these two genera are also able to produce flashes of light when disturbed; *Plocamopherus* has light organs on some of the lateral papillae.

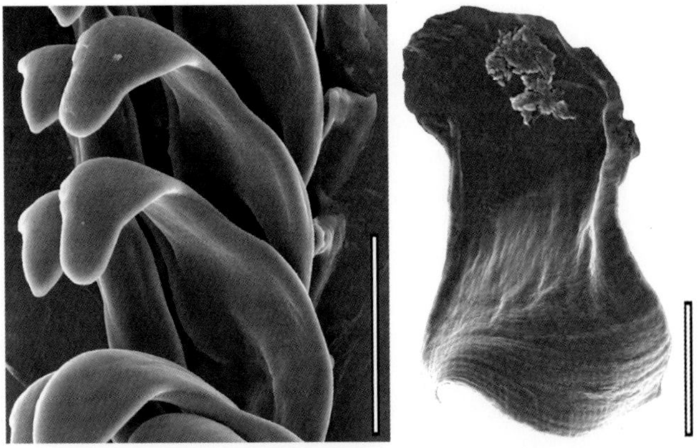

photos A. Valdés - SEM of *Polycera herthae*; left: radular teeth, scale bar = 50 μm; right: jaw, scale bar = 100 μm

Polycera hummi Abbott 1952: 3-4, pl. 2, figs. 1-8.

Description: Background color translucent brownish gray, covered with a close scattering of black or dark brown specks. Body with small conical papillae. Papillae, gill, and rhinophores, translucent gray-brown. Three extra-branchial and four velar appendages with three characteristic color bands, a sub-terminal dark blue band, followed by a yellow-orange band and a light blue band about midway down each papilla.

Geographic Range: North Carolina, Florida.

Natural History: Found on the arborescent bryozoan *Bugula.*

Polycera herthae Ev. Marcus & Er. Marcus 1963: 34-35, figs. 43-46.
Polycera manzanilloensis Ortea, Espinosa & Camacho 1999: 161-163, fig. 3.

Description: Background color from pale to very dark gray with many shades of brown. Rhinophores and gill with the same color as the body and white to yellow apices. Sides of the body with small conical papillae with a translucent base and a yellow, rounded apex. Velum with eight to ten finger-like papillae, white and yellow. Dorsum with two tubercles or concentrations of tubercles, one behind the rhinophores and one in front of the gill. Gill elevated, situated on a dorsal hump.

Geographic Range: Florida, Costa Rica, Cuba, Curaçao, Antigua, Bahamas.

Remarks: *Polycera manzanilloensis* seems to be based on juvenile specimens of *P. herthae.*

Polycera rycia
Er. Marcus & Ev. Marcus, 1970

"Papillose polycera"

Size: Up to 9 mm long

photo © L. Ianniello - Riviera Beach, Florida

Polycera chilluna
Er. Marcus, 1961

Size: Up to 25 mm long

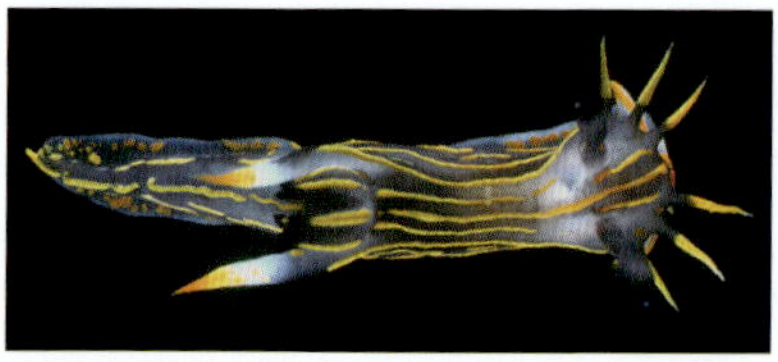

photos J. Hamann, © E. Ruppert & R. Fox - left: Venezuela (JH); right: North Carolina (ER & RF)

Polycera hedgpethi
Er. Marcus, 1964

Size: Up to 25 mm long

photo J. Hamann - California

Polycera rycia Er. Marcus & Ev. Marcus 1970: 71-72, figs. 125-126.

Description: Background color light gray mottled all over with brown spots and covered with small white papillae. Velar processes light blue, with an orange band near the base, a dark blue middle zone and a light blue tip. Four extra-branchial appendages, two on each side, with a blue ring in the middle followed by a yellowish orange band and a light tip; anterior pair smaller, club-shaped, posterior pair much larger, with small papillae.

Geographic Range: Florida, Bahamas.

Remarks: The Brazilian species *Polycera aurisula* differs from *Polycera rycia* by having three extra-branchial processes on each side, all of them club-shaped with no papillae.

Polycera chilluna Er. Marcus 1961: 143-144, figs. 6-9.

Description: Background color variable from translucent light to dark gray or black. In dark specimens more translucent around the anterior end. Several longitudinal yellow lines run from in front of the rhinophores to the gill, as well as on the sides and posterior end of the body. Rhinophores and gill translucent with dark apices. Velar and extra-branchial processes translucent with yellow tips.

Geographic Range: North Carolina, Venezuela.

Remarks: The eastern Atlantic species *Polycera aurantiomarginata* García Gómez & Bobo, 1984 could be a synonym.

Polycera hedgpethi Er. Marcus 1964: 128-131, figs. 1-4.

Description: Background color translucent gray covered with black or gray speckling. Velum with 4-6 pointed yellow processes with a band of gray-black and white tips. Mantle margin and center of the posterior end of the foot white to yellow. Rhinophores gray with a black band and a yellow apex. Branchial leaves gray or blackish with yellow tips. Extra-branchial appendages similar in color to the anterior frontal veil processes.

Geographic Range: Cosmopolitan (probably introduced in many localities); Western Atlantic: Costa Rica.

Natural History: Member of the fouling community associated with the bryozoan *Bugula*.

Polycera odhneri
Er. Marcus, 1955

"Odhner's polycera"

Size: Up to 25 mm long

photos J. Hamann - left: Venezuela; right: Bermuda

photo J. Hamann - St. Vincent & the Grenadines

Polycera incognita
(Ortea, Espinosa & Caballer, 2004)

Size: Up to 7 mm long

photo J. Hamann - Cozumel, Mexico

Polycera sp. 1

Size: Up to 7 mm long

Polycera odhneri Er. Marcus 1955: 158-161, fig. 193-202.

Description: Background color of this animal varies from light brown to black-brown, sometimes with a darker area at the top of the flanks. Most of the upper surface, including the gills and rhinophores, is covered with small spots of brick-red or orange, with a scattering of much smaller yellow dots.

Geographic Range: Florida, Venezuela, Curaçao, Cuba, Barbados, Bahamas, Bermuda, Brazil.

Natural History: Found on floating *Sargassum* and on a variety of algae.

Kankelibranchus incognitus Ortea, Espinosa & Caballer 2004: 103-105, figs. 1-2.

Description: Background color reddish brown with numerous black spots. Velar appendages and lateral papillae whitish with numerous small white branches. Similar appendages occur on the sides and posterior end of the foot. Rhinophores and gill orange with some white pigment.

Geographic Range: Cuba, Bahamas, Barbados, St. Vincent & the Grenadines.

Remarks: There is little phylogenetic justification for the introduction of a new genus for this species, therefore *Kankelibranchus* is here regarded as a synonym of *Polycera.*

Description: Background color dark brown with a few opaque white spots. Mantle margin and posterior end of the foot with a series of translucent conical papillae with a yellow apex. Rhinophores uniformly brown; gill brown with white apices.

Geographic Range: Mexico.

Remarks: This appears to be an undescribed species.

photo J. Hamann - Laughing Bird Cay, Belize

Polycera sp. 2

Size: Up to 16 mm long

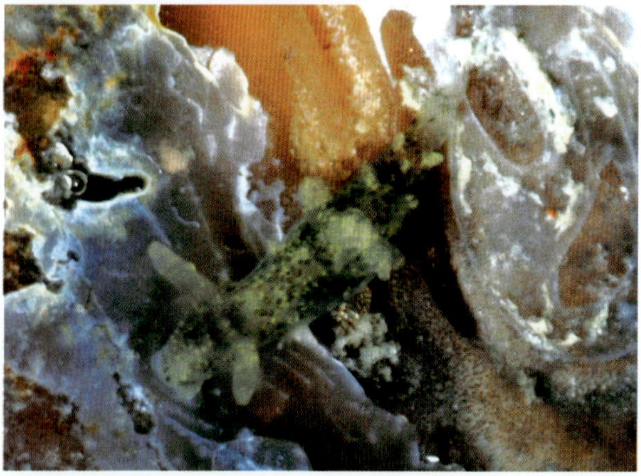

photo A. Valdés - Captiva Island, Florida

Polycerella emertoni

Verrill, 1880

"Atlantic polycerella"

Size: Up to 6 mm long

photo J. Hamann - Panama

Tambja oliva

Meyer, 1977

Size: Up to 20 mm long

Description: Background color translucent gray with the viscera visible through as a red mass. Dorsum and sides covered by a series of reddish longitudinal lines. Body with a number of conical, simple yellow papillae surrounded with opaque white pigment and with a rounded white apex. Velar appendages and lateral papillae reddish with numerous small white branches. Similar appendages occur on the sides and posterior end of the foot. Rhinophores and gill translucent with red pigment, white spots and white apices.

Geographic Range: Florida, Belize, Panama.

Remarks: This appears to be an undescribed species.

Polycerella emertoni Verrill 1880a: 387-388.
Polycerella davenportii Balch 1899: 150-151, figs. 2-7.
Polycerella conyna Er. Marcus 1957: 429-432, figs. 104-110.

Description: Background color translucent grayish white, with the brownish viscera showing through in the central area. Numerous flecks and blotches of dark green and brown present in some animals. Opaque white dots coalesce into small random patches, but mostly concentrated on the rhinophores and on a series of small papillae that line the dorsal ridge. Two large appendages behind the three translucent branchial leaves.

Geographic Range: Massachusetts, Rhode Island, Connecticut, New York, New Jersey, Virginia, Florida, Venezuela, Brazil.

Natural History: Feeds on arborescent bryozoans *Zoobotryon* sp, *Bowerbankia gracilis*, *Amathia distans* and *Bugula neritina*.

Tambja oliva Meyer 1977: 302-304, figs. 1-2.

Description: Background color deep olive green to light green or olive yellow, often with microscopic flecks of yellow, blue or turquoise. Tips of the branchial leaves, posterior foot, and oral tentacles light to dark purple or a dark blue-green. Rhinophores very dark, often purple or translucent white with opaque cream spots.

Geographic Range: Costa Rica, Panama.

photo © P. Humann - Venezuela

Tambja* cf. *stegosauriformis
Pola, Cervera & Gosliner, 2005

Size: Up to 50 mm long

photo © H. Segars - South Carolina

Roboastra gratiosa
(Bergh, 1890)

Size: Up to 22 mm long

photo © K. Johnson - Florida

Thecacera pennigera
(Montagu, 1815)

Size: Up to 20 mm long

Tambja stegosauriformis Pola, Cervera & Gosliner 2005: 979-983, figs. 1, 2A-F, 3A.

Description: Background color dark brown, with large, elongate green spots with bluish green edges. Head with a characteristic frontal green extension edged with black. Rhinophores with a brown base and black tips. Branchial leaves orange with black tips.

Geographic Range: Venezuela, Brazil.

Remarks: The record from Venezuela here illustrated has a color pattern somewhat different from the original specimens from Brazil, thus this identification needs to be confirmed using anatomical data.

Tambja gratiosa Bergh 1890: 172-175, pl. 2, figs. 1-5, pl. 3, figs. 1-4.

Description: Background color bright yellow, with large, rounded black spots with faded edges. A network of light yellow shallow grooves extends over the dorsum and sides. Rhinophores and branchial leaves with yellowish bases and deep purple apices. Anterior end of the head with a black edge; oral tentacles long, deep purple.

Geographic Range: South Carolina, Florida, Dry Tortugas, Bahamas.

Remarks: According to Pola *et al.* (personal communication), this species is a member of the genus *Roboastra*.

Doris pennigera Montagu 1815: 17.
Thecacera maculata Eliot 1905: 241-242.
Thecacera lamellata Barnard 1933: 294-295, fig. 1.

Description: Background color translucent gray, almost transparent, covered with yellow and black spots and some opaque white flecks. Rhinophores and gill with the same color. Two pairs of tentacular appendages, one pair on each side of the rhinophores and one pair on each side of the gill, with the same color as the body but with an opaque white gland in the apex. Oral tentacles conical, grooved.

Geographic Range: Cosmopolitan; Western Atlantic: Florida, Brazil.

Natural History: Feeds on the arborescent bryozoan *Bugula*.

Thecacera pacifica
(Bergh, 1883)

Size: Up to 30 mm long

photo © J. Burek - Flower Garden banks, Texas

Plocamopherus lucayensis
Hamann & Farmer, 1988

Size: Up to 40 mm long

photos © J. Worsfold & J. Hamann - left: Bahamas (JW); right: spotted color form, Curaçao (JH)

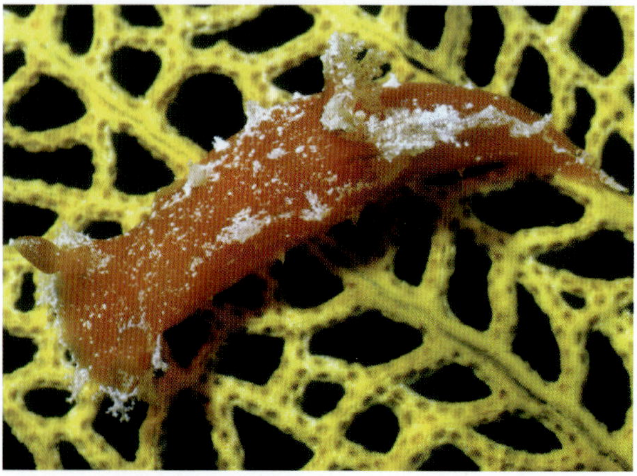

Plocamopherus pilatectus
Hamann & Farmer, 1988

Size: Up to 28 mm long

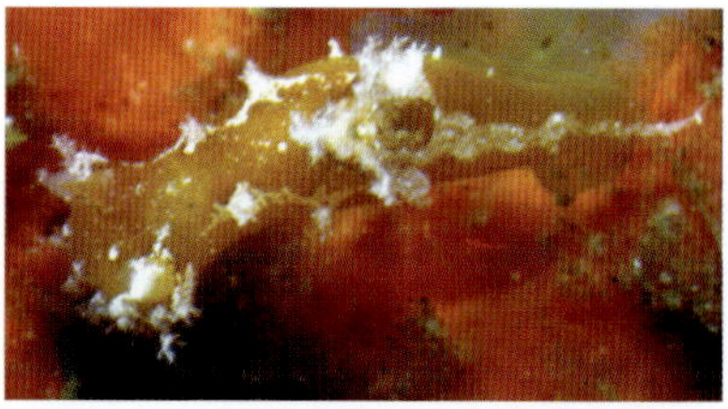

photo J. Hamann - left: Bequia, St. Vincent & the Grenadines; right: Guadeloupe, juvenile

Ohola pacifica Bergh 1883: 136-137.

Description: Background color uniformly reddish brown. Rhinophores and gill with the same color as the dorsum, black tips on the rhinophores and black rachises on the branchial leaves. Two pairs of tentacular appendages on the sides of the rhinophores and gill, with the same color as the body but with the apex black to bright metallic blue. A spot with a similar color at the posterior end of the body. Rhinophoral sheaths with a black edge.

Geographic Range: Indo-Pacific; Western Atlantic: Texas.

Plocamopherus lucayensis Hamann & Farmer 1988: 68-71, figs. 1-6.

Description: Background color bright red to brown, with numerous small opaque white papillae all over the body with a black ring at the base. Some specimens with rows of white spots on the dorsum. Rhinophores uniformly red. Branchial leaves red with black and white apices. Velar appendages slightly branched, with a red base, black ring at mid-length and a white apex. A larger appendage on each side of the gill bearing a large, white, oval light organ at the tip. Posterior end of the foot wide with white papillae.

Geographic Range: Bahamas.

Natural History: This species produces flashes of light when disturbed.

Plocamopherus pilatecta Hamann & Farmer 1988: 71-74, figs. 7-12.

Description: Background color pale reddish gray, with abundant opaque white pigment all over the body. Rhinophores reddish with white spots. Branchial leaves reddish with white apices. Velar appendages highly branched, white. A few conical white appendages on each side of the body, the posterior ones bearing a large, white, oval light organ at the tip. Posterior end of the foot wide with a longitudinal white ridge and some white papillae.

Geographic Range: Cayman Islands, Guadeloupe, St. Vincent & the Grenadines.

Natural History: This species produces flashes of light when disturbed. It can also swim by lateral flexions of the posterior end of the body.

Family CHROMODORIDIDAE Bergh, 1891

Diagnosis: Body oval to elongate, generally with a wide mantle margin, very elaborate in some groups. Most species are brightly colored. Body surface usually soft, without spicules, tubercles or papillae, although these are present in some groups. Defensive mantle glands of different shapes well developed and arranged in a characteristic way in each group. Rhinophores and gill retract into pockets; gill with unbranched leaves. Reproductive system with a vestibular gland in most groups, in some cases large and ramifying. Radular teeth numerous and strongly denticulate. Some groups have a reduced rachidian tooth. Jaw plates present in most groups, consisting of tightly packed clusters of rodlets, sometimes with bifid or multifid cusps.

Biology: Most species feed on non-spiculate sponges that produce toxic chemicals for defense against predators. The nudibranchs ingest these chemicals and accumulate them in their mantle glands in order to use them for their own defense. Chromodorids have a wide range of reproduc-

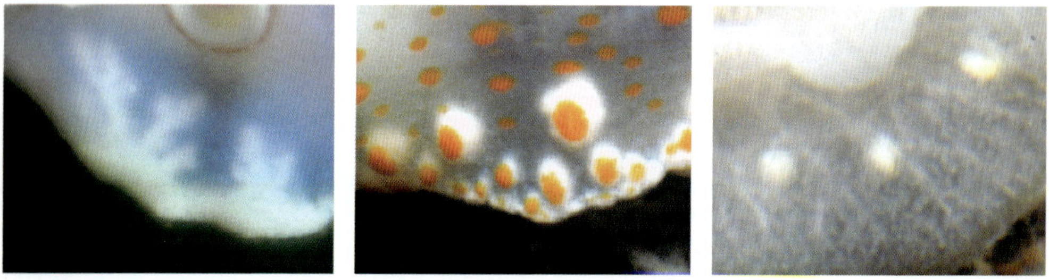

photos J. Hamann - Different types and arrangements of mantle glands; left: ramifying glands arranged in one row in *Chromodoris* sp.; center: simple glands arranged in two rows in *Tyrinna evelinae*; right: simple glands arranged in one row in *Cadlina rumia*.

photo J. Hamann - Samaná, Dominican Republic

Tyrinna evelinae
(Er. Marcus, 1958)

"Eveline's tyrinna"

Size: Up to 30 mm long

tive strategies from having relatively long-term, planktotrophic veliger larvae to being short-term direct developers. Direct developing species usually have a reduced geographic range.

Remarks: The taxonomy of several Caribbean species of Chromodorididae is in need of review. The following species have not been collected since their original description and their identity remains uncertain: *Chromodoris aila* Er. Marcus, 1961; *Chromodoris dictya* Er. Marcus & Ev. Marcus, 1970; *Chromodoris perola* Ev. Marcus, 1976; and *Chromodoris moerchi* Bergh, 1879.

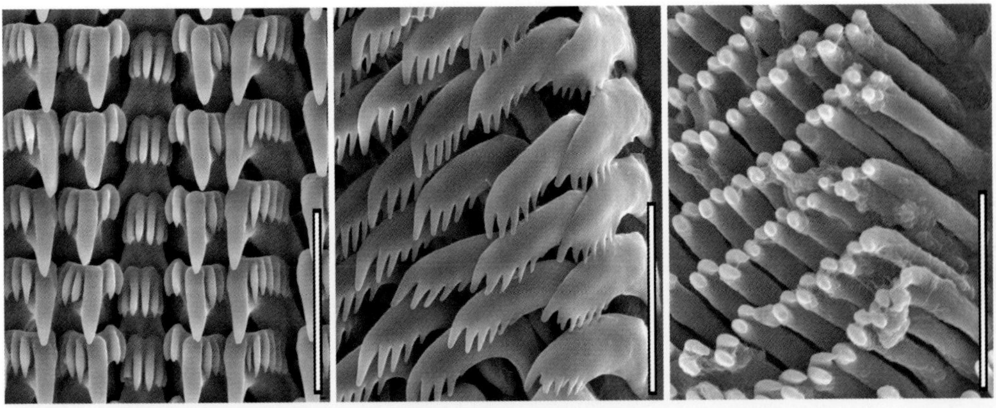

photos A. Valdés - SEM of *Cadlina rumia*; left: central radular teeth, scale bar = 30 μm; center: outer radular teeth, scale bar = 30 μm; right: jaw rodlets, scale bar = 20 μm

---

Cadlina evelinae Er. Marcus 1958: 18-21, figs. 28-33.
Cadlina burnayi Ortea 1988: 2-5, figs. 1-2, pl. 1A.

Description: Background color translucent white with a number of orange spots. Rhinophores and branchial leaves translucent white with opaque white tips. Mantle margin edged by an opaque white line. Mantle glands white with orange centers arranged in two rows.

Geographic Range: Eastern Atlantic; Eastern Pacific; Western Atlantic: Costa Rica, Jamaica, Puerto Rico, Dominican Republic, Brazil.

Natural History: Intertidal to 5 m depth.

Remarks: Caribbean populations are anatomically indistinguishable from those in the Eastern Pacific and Eastern Atlantic, therefore they are all included in the same species.

Cadlina rumia
Er. Marcus, 1955

"Pale cadlina"

Up to 12 mm long

photo J. Hamann - Trujillo, Honduras

Glossodoris punctilucens
Bergh, 1890

Up to 35 mm long

photo © S. Kaicher - Florida

Glossodoris sedna
(Ev. Marcus & Er. Marcus, 1967)

"Red-tipped sea goddess"

Up to 65 mm long

photo J. Hamann - Pelican Keys, Belize

Cadlina rumia Er. Marcus 1955: 119-124, figs. 78-89.

Description: Background color from white to creamy yellow. Rhinophores sometimes as pale as the body, but can also have a darker, golden brown tint. Mantle glands yellow, forming a single line around the mantle. Branchial leaves sometimes slightly darker than the body.

Geographic Range: Florida, Belize, Panama, Bahamas, Dominican Republic, Jamaica, Puerto Rico, Curaçao, St. Maarten/St. Martin, St. Lucia, St. Vincent & the Grenadines, Grenada, Brazil.

Natural History: Intertidal.

Remarks: This is the only species of *Cadlina* found in the Western Atlantic.

Glossodoris punctilucens Bergh 1890: 162-165, pl. 1, figs. 4-10.

Description: Background color dark blue to black with numerous white spots all over the dorsum. Mantle margin edged by a bluish marginal band, followed by a white line, a black area and a whitish band formed by the mantle glands. Rhinophores and gill dark with white apices.

Geographic Range: Florida, Dry Tortugas.

Casella sedna Ev. Marcus & Er. Marcus 1967a: 178-181, figs. 34-37.

Description: Background color translucent white with both foot and mantle bordered with three color bands; an inner opaque white, followed by a red and an outer yellow band. Upper half of rhinophoral clubs and tips of branchial leaves are red.

Geographic Range: Eastern Pacific and Western Atlantic: Florida, Belize, Bahamas.

Natural History: Inhabits reefs and rocky areas, feeds on sponges.

Remarks: The Caribbean population is probably the result of a recent introduction.

Chromodoris clenchi
(Russell, 1935)

"Harlequin blue sea goddess"

Size: Up to 30 mm long

photo J. Hamann - Piton Bay, St. Lucia

photo A. Valdés - Costa Rica, juvenile specimen

photo J. Hamann - Cane Garden Bay, Tortola (specimen with egg mass)

Chromodoris binza
Ev. Marcus & Er. Marcus, 1963

Size: Up to 30 mm long

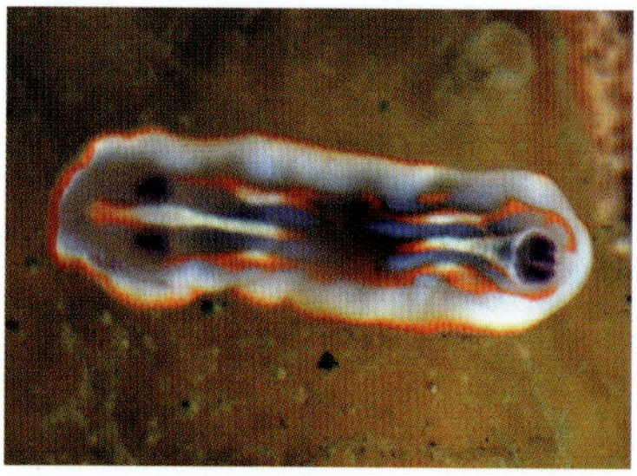

photos J. Hamann & A. Valdés - left: St. Vincent & the Grenadines (JH); right: Costa Rica (AV)

Glossodoris clenchi Russell 1935: 59, fig. 5.

Description: Background color pale blue, with a dense pattern of red covering the entire dorsum and leaving small, circular or oval uncovered areas. The red becomes yellow near the rhinophores and the gill. Mantle margin with a submarginal white band partially covered with extensions of the red pattern. Mantle edge with a red line. Rhinophores and gill pale whitish gray with purple rachises. Juvenile specimens with a less dense red pattern.

Geographic Range: Florida, Costa Rica, Panama, Colombia, Bermuda, Cayman Islands, Jamaica, Curaçao, St. Lucia, St. Vincent & the Grenadines.

Remarks: This species has been often confused with *Chromodoris neona* and *Chromodoris binza*, which are distinct species (see next page).

photos J. Hamann - Sewage Beach, Florida (rhinophores - left; gill - right)

Chromodoris binza Ev. Marcus & Er. Marcus 1963: 25-26, figs. 30-31.

Description: Background color pale blue, with a reticulate pattern of red covering parts of the dorsum and leaving large uncovered areas. The red becomes yellow near the rhinophores and the gill. Mantle margin with a submarginal white band partially covered with extensions of the red pattern. Mantle edge with a red line. Rhinophores blue with longitudinal purple bands. Gill pale whitish gray with purple rachises. Juvenile specimens with a less dense red pattern.

Geographic Range: Florida, Costa Rica, Colombia, Cuba, Virgin Islands, Curaçao, St. Vincent & the Grenadines, Brazil.

Remarks: Very similar to *Chromodoris clenchi*, these two species are often confused. The main differences are the less dense pattern of red and darker rhinophores in *C. binza*. The authors of this book do not agree on whether these two species are synonyms, but we prefer to leave them separate until new studies are conducted.

Chromodoris* cf. *ponga
Er. Marcus & Ev. Marcus, 1970

Size: Up to 10 mm long

photo © H. Bertsch - Panama

Chromodoris neona
(Er. Marcus, 1955)

"Neon sea goddess"

Size: Up to 30 mm long

photo J. Hamann - Sewage Beach, Florida

photos J. Hamann - Sewage Beach, Florida; left: specimen with egg mass; right: juvenile specimen

Chromodoris ponga Er. Marcus & Ev. Marcus 1970: 55-56, figs. 99-101.

Description: Background color white, with a large patch of red covering most of the dorsum leaving two uncovered areas in front of the gill and behind the rhinophores. A few purple spots scattered over the dorsum. Mantle margin with a submarginal white band partially covered with extensions of the red patch. Mantle edge with a red line surrounded by a gray line. Rhinophores and gill pale whitish gray with purple rachises.

Geographic Range: Curaçao, Panama.

Remarks: This species is most likely a color form of *Chromodoris clenchi* and along with *Chromodoris binza* anatomical studies are necessary to clarify its status.

Glossodoris neona Er. Marcus 1955: 124-127, figs. 90-101.

Description: Background color pale bluish gray with a pattern of reticulate yellow lines edged with a reddish shade. Mantle margin with a white band partially covered with extensions of the yellow pattern. Mantle edge with a yellow submarginal line and a gray marginal line. Rhinophores and gill white with blue tips.

Geographic Range: Florida, Panama, Brazil.

Remarks: This species has often been confused with *Chromodoris clenchi* and *C. binza* and regarded as a synonym by some authors. The main differences between these three species is the presence of white rhinophores and branchial leaves with blue tips and the characteristic pattern of yellow lines on the dorsum of *C. neona*.

photos J. Hamann - Sewage Beach, Florida, left: rhinophores and oral tentacles; right: gill

Chromodoris grahami
Thompson, 1980

Size: Up to 20 mm long

photos J. Hamann - left: Kingstown, St. Vincent; right: juvenile, Chimana Grande, Venezuela

Chromodoris regalis
(Ortea, Caballer, & Moro, 2001)

Size: Up to 34 mm long

photos J. Hamann - left: Trou Madame, Martinique; right: juvenile, St. Vincent & the Grenadines

Chromodoris sp.

Size: Up to 8 mm long

photo J. Hamann - Isla Mujeres, Mexico

Chromodoris grahami Thompson 1980: 80-81, figs. 4a, 5c.

Description: Background color varies from white to salmon pink with bright red spots. There is a broad marginal white band with a yellow submarginal line at the mantle edge. Foot with a broad white border and a white line down the posterior mid-line. Rhinophores dark brown, speckled with white and having white apices. Branchial leaves white with purple tips.

Geographic Range: Panama, Venezuela, Jamaica, Puerto Rico, St. Lucia, St. Vincent & the Grenadines, St. Maarten/St. Martin.

Natural History: Found from 1-5 m depth. Hides under coral.

Noumea regalis Ortea, Caballer, & Moro 2001: 2-6, figs. 1-3.

Description: Background color salmon pink with numerous opaque white spots. There is a broad sub-marginal white band and a yellow line right at the mantle edge. Foot with a broad white border. Rhinophores salmon pink. Branchial leaves pink with red rachises.

Geographic Range: Costa Rica, Martinique, St. Vincent & the Grenadines.

Natural History: Found on rocky bottoms from 2-9 m depth.

Remarks: Originally described as a member of the Indo-Pacific genus *Noumea*, the anatomical features of this species are characteristic of *Chromodoris*.

Description: Background color pale blue with a pinkish tinge and a series of large, white, oval spots on the dorsum surrounded by a pinkish line. Four larger, irregular spots occur behind the rhinophores and in front of the gill, the latter are fused together around the gill opening. Rhinophores and gill pinkish with the base white. Mantle margin surrounded by a yellow line. Mantle glands ramifying, opaque white.

Geographic Range: Mexico.

Natural History: Found at 10 m depth.

Remarks: This appears to be an undescribed species.

photo J. Hamann - Long Point, Bahamas

photo J. Hamann - Cap Salomon, Martinique

Risbecia nyalya
(Ev. Marcus & Er. Marcus, 1967)

"Red line blue sea goddess"

Size: Up to 35 mm long

photo J. Hamann - Florida

Hypselodoris picta
(Schultz, 1836)

"Florida regal doris"

Size: Up to 130 mm long

Chromodoris nyalya Ev. Marcus & Er. Marcus 1967a: 53-55, figs. 62-66, pl. 1, fig. 1.

Description: Background color varies from whitish blue, to light blue to a darker blue. Mantle bordered with a red band. Rhinophores and gill blue. Dorsal side of the posterior end of the foot blue with no markings.

Geographic Range: Florida, Bahamas, Cuba, Martinique.

Natural History: Inhabits reefs, prefers areas of moderate current. Found on sponges of the genus *Dysidea*. The large size of the eggs indicates that this species could have direct development.

Remarks: Recently transferred to the genus *Risbecia*.

photo J. Hamann - Long Point, Bahamas; left: egg mass; right: adult specimen

---

Doris elegans Cantraine 1835: 383 [non *Doris elegans* Quoy & Gaimard, 1832]
Doris picta Schultz *in* Philippi 1836: 105.
Glossodoris edenticulata White 1952: 113-114, figs. 13-15, pl. 6, fig 3.
[other synonyms not included]

Description: Background color dark blue with a series of longitudinal yellow lines and yellow spots on the dorsum. Rhinophores uniformly blue. Gill with yellow rachises.

Geographic Range: Widespread in the Atlantic Ocean; Western Atlantic: Florida, Brazil.

Natural History: Intertidal to 50 m depth. In the Mediterranean feeds on *Dysidea fragilis*.

Remarks: Very variable species found on both coasts of the Atlantic. Several subspecies have been recognized based on regional color patterns.

Hypselodoris zebra
Heilprin, 1888

"Zebra slug"

Size: Up to 115 mm long

photo J. Hamann - Harrington Sound, Bermuda

Hypselodoris ruthae
Ev. Marcus & Hughes, 1974

"Gold line sea goddess"

Size: Up to 27 mm long

photo J. Hamann - Cancún, Mexico

photos J. Hamann - left: Sandy Cay, Virgin Islands; right: head, Chocolate Bay, Virgin Islands

Hypselodoris zebra Heilprin 1888: 327, pl. 16, figs. 3-3A.
?*Chromodoris roseopicta* A. E. Verrill 1900: 549, pl. 66, fig. 1.

Description: Background color dark blue, with a reticulate pattern of white, yellow lines, mostly running longitudinally. Rhinophores uniformly dark blue to purple, branchial leaves grey with dark rachises and yellow longitudinal lines forming inverted triangles in the largest specimens. A wide marginal band of white or grey with a series of black spots surrounds the mantle margin. Edge of the mantle with a thin greyish orange line.

Geographic Range: Bermuda.

Remarks: *Hypselodoris sycilla* (Bergh, 1890) could be a synonym, so this species would have a wider geographic range in the Caribbean including Florida, Mexico, Puerto Rico and Brazil.

Hypselodoris ruthae Ev. Marcus & Hughes 1974: 518-530, figs. 36-40.

Description: Background color dark blue. Mantle with a marginal yellow border surrounding a broad white band, which is interrupted in the middle of the body by blue pigment. There are up to 5-6 longitudinal yellow lines on the dorsum, the number of lines increasing with the size of the animal. Rhinophores blue and branchial leaves white edged with blue lines.

Geographic Range: Mexico, Costa Rica, Venezuela, Bahamas, Cuba, Jamaica, Puerto Rico, Virgin Islands, Aruba, Curaçao, St. Maarten/St. Martin, Antigua, Guadeloupe, Martinique, St. Lucia, Barbados, Grenada.

Natural History: Found from 1-12 m depth. Inhabits living reef, feeds on blue sponge *Dysidae janiae.*

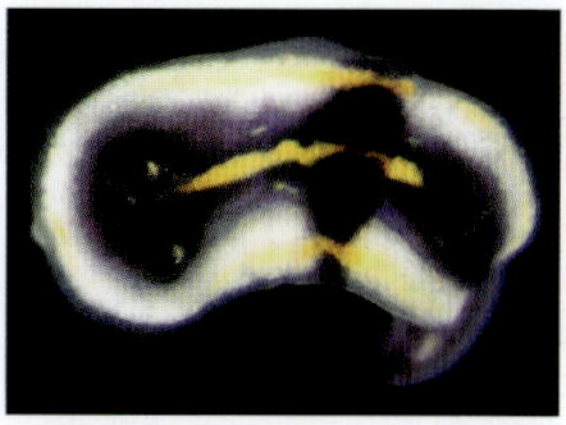

photo © L. Harris - juvenile specimen, Virgin Islands

photos A. Valdés - left: lateral view, Puerto Viejo, Costa Rica; right: juvenile specimens, Curaçao

Hypselodoris bayeri
(Ev. Marcus & Er. Marcus, 1967)

"Black-spotted sea goddess"

Size: Up to 60 mm long

photo J. Hamann - Biscayne Bay, Florida

photos J. Hamann - left: Arrecife Alacrán, Mexico; right: ventral view of head, Biscayne Bay, Florida

Hypselodoris marci
Ev. Marcus, 1971

"Purple-spotted sea goddess"

Size: Up to 26 mm long

photos J. Hamann - Lark Key, Belize; left: adult specimen; right: rhinophores and oral tentacles

Felimare bayeri Ev. Marcus & Er. Marcus 1967a: 62-64, figs. 76-79, pl. 1, fig. 2.

Description: Background color dark blue, with a series of longitudinal yellow lines. Rhinophores uniform purple-blue; branchial leaves opaque white with dark rachises. A wide submarginal band of white with a series of black circular spots around the mantle margin. Edge of the mantle with a wide yellow line. Posterior end of the foot blue with yellow lines and black spots.

Geographic Range: Florida, Panama, Cuba, Mexico, Belize.

Remarks: Originally described as the only member of the new genus *Felimare*, it was subsequently transferred to *Hypselodoris* of which *Felimare* is a synonym.

photos J. Hamann - Arrecife Alacrán, Mexico; left: rhinophores; right: gill

---

Hypselodoris marci Ev. Marcus 1971: 941-943, figs. 39-43.

Description: Background color blue, with a reticulate pattern of white, yellow or orange. Rhinophores white with purple-blue tips, branchial leaves white with dark rachises. A wide marginal band of white with a series of large black spots borders the mantle. Edge of the mantle with a thin greyish orange line.

Geographic Range: Belize, Venezuela, Brazil.

Remarks: *Hypselodoris marci* was described as having a reticulate pattern of yellow, large black spots on the mantle margin, and rhinophores with blue tips. Other specimens assigned to *H. marci* by Ev. Marcus and other authors probably are *Hypselodoris* sp. 3.

photo © J. Holdorf - Honduras

photo D. Behrens - Roatán, Honduras

Hypselodoris acriba

Ev. Marcus & Er. Marcus, 1967

"Gold-crowned sea goddess"

Size: Up to 60 mm long

photo J. Hamann - juvenile, Vieux Fort Bay, St. Lucia

photo J. Hamann - Vieux Fort Bay, St. Lucia

photo J. Hamann - Pt. Falaise, St. Maarten/St. Martin

Hypselodoris cf. *acriba*

Several specimens illustrated here are considered color variations of *H. acriba* but anatomical examinations are necessary to confirm this point. They are characterized by having small rounded or elongate black spots on the white band around the mantle margin and not only on the inner edge of this band (left: specimen with small black spots; right: specimen with longitudinal black stripes).

Hypselodoris acriba Ev. Marcus & Er. Marcus 1967a: 60-62, figs. 72-75.

Description: Background color navy blue, covered with a fine yellow-gold mesh pattern. Rhinophores uniform purple-blue and branchial leaves white to gold with dark rachises. A wide marginal band of white to pink lies inside of the highly ruffled mantle margin. Edge of the mantle with a thin yellow line followed by a red line. A series of irregular black spots occurs along the inner edge of the ruffled margin. The yellow mesh network occurs over the posterior end of the foot.

Geographic Range: Mexico, Costa Rica, St. Lucia, Guadeloupe, St. Martin, Puerto Rico.

Natural History: Inhabits reefs, feeds on sponges.

Remarks: *Hypselodoris acriba* is distinguishable from *Hypselodoris bayeri* by having a reticulate pattern of yellow instead of longitudinal lines. Differences with *H. marci* include the uniformly purple rinophores and the absence of large black circular spots on the mantle margin.

photos J. Hamann - Terre d'en Haut, Guadeloupe; left: oral tentacles; right: rhinophores

photos J. Hamann - Esperanza, Vieques, Puerto Rico; left: adult specimen; right: oral tentacles

Hypselodoris espinosai

Ortea & Valdés, 1996

"Espinosa's sea goddess"

Size: Up to 17 mm long

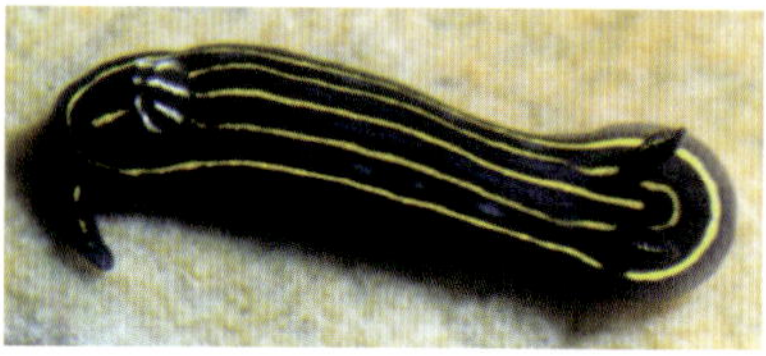

photos J. Hamann - left: Marigot, St. Martin; right: Union Island, St. Vincent & the Grenadines

Hypselodoris lilyeveae

Alejandrino & Valdés, 2006

"Lily Eve's sea goddess"

Size: Up to 26 mm long

photo J. Hamann - Rhoda Rocks, Bahamas

photos J. Hamann - Luis Peña Cay, Puerto Rico; left: dorsal view; right: laying eggs

Hypselodoris espinosai Ortea & Valdés *in* Ortea *et al.* 1996: 139-142, figs 12E, 109-111.

Description: Background color dark blue with three longitudinal bright yellow lines that are joined in front of the rhinophores and behind the gill. Mantle margin with a submarginal yellow line. There are several large light blue spots on the dorsum. Rhinophores and gill blue with white rachises.

Geographic Range: Mexico, St. Martin, St. Vincent & the Grenadines.

Natural History: Found at 4 m depth.

Hypselodoris lilyeveae Alejandrino & Valdés 2005: 191-192, figs. 1-3.

Description: Background color orange-red, fading into grayish towards the center of the dorsum. A broad, irregular, longitudinal white band runs from the rhinophores to the posterior end of the body. Dorsum with a number of small black spots and a series of bright blue spots forming two lines on the lateral edges of the mantle. Rhinophores white, with three purple longitudinal bands. Branchial leaves white with a purple rachis.

Geographic Range: Puerto Rico, Virgin Islands, Bahamas.

Natural History: Found from 12-18 depth.

Remarks: This is the only Atlantic species of *Hypselodoris* with a reddish background color. The large size of the eggs of this species indicates that it is probably a direct developer.

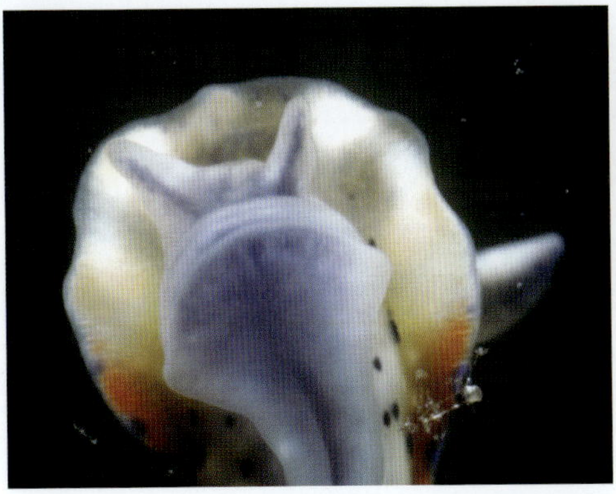

photos J. Hamann - Luis Peña Cay, Puerto Rico; left: ventral view of head; right: egg mass

Hypselodoris sp. 1

Size: Up to 50 mm long

photos © H. Bertsch & A. DuPont - left: Grenada (JH); right: Florida (AD)

Hypselodoris sp. 2

Size: Up to 40 mm long

photo J. Hamann - St. Vincent & the Grenadines

Hypselodoris sp. 3

Size: Up to 40 mm long

photos J. Hamann - Luis Peña Canal, Puerto Rico; left: dorsal view; right: specimen feeding

Description: Background color pale yellow to pale blue with a number of longitudinal yellow lines covering the entire center of the dorsum except for the areas around the eyes. A wide submarginal band of white lies inside the mantle margin, interrupted by a dark area near the center of the animal. Edge with a yellowish red line. Rhinophores dark blue with a white rachis. Branchial leaves branched, white with dark rachises.

Geographic Range: Florida, Grenada.

Remarks: This appears to be an undescribed species.

Description: Background color dark blue to black with a number of longitudinal yellow lines covering the entire center of the dorsum except for the areas around the eyes. A thin yellow line around the mantle margin, edged with an opaque white line. Rhinophores uniformly purple. Branchial leaves grey with dark gray and yellow rachises.

Geographic Range: St. Vincent & the Grenadines.

Remarks: This appears to be an undescribed species.

photos J. Hamann - St. Vincent & the Grenadines; left: head; right: gill

---

Description: Background color white with irregular shades of pale blue and gray. Dorsum with three longitudinal yellow lines becoming orange near the center of the body. Rhinophores white with a purple-blue longitudinal line; branchial leaves opaque white with rachises white basally and dark gray apically. Mantle margin white, with a narrow submarginal band of yellow becoming orange near the center, and a series of black circular spots.

Geographic Range: Venezuela, Puerto Rico, Virgin Islands, Curaçao.

Natural history: Feeds on a blue sponge of the genus *Dysidea*.

photos J. Hamann - left: oral tentacles, Puerto Rico; right: egg mass, Virgin Islands

Remarks: This appears to be an undescribed species. The external coloration resembles that of *Hypselodoris bayeri*, but the rhinophores are white with longitudinal purple bands.

photo © L. Ianniello - Deerfield Beach, Florida

Hypselodoris sp. 4

Size: Up to 60 mm long

photo J. Hamann - Cancún, Mexico

Mexichromis kempfi
(Ev. Marcus, 1971)

“Purple crowned sea goddess”

Size: Up to 20 mm long

photos A. Valdés - Puerto Viejo, Costa Rica; left: adult specimen; right: juvenile specimen

Description: Background color yellowish cream, with shades of ochre brown and numerous bright blue, irregular spots surrounded by black dots. Mantle margin with numerous black dots. Rhinophores white with purple apices. Branchial leaves opaque white with black rachises.

Geographic Range: Florida.

Natural History: Feeds on the blue sponge *Dysidea etheria*.

Remarks: This species has been confused with *Hypselodoris marci*, but it is probably undescribed.

Chromodoris kempfi Ev. Marcus 1971: 940-941, figs. 34-38.
?*Mexichromis molloi* Ortea & Valdés *in* Ortea *et al.* 1996: 152-155, figs. 12F, 118-120.

Description: Background color opaque white, with a bluish central area surrounding a longitudinal opaque white band that runs from between the rhinophores to the gill. There is a series of elongate black spots on each side of the medial white band. Mantle margin with a broad orange band. Rhinophores entirely purple. Branchial leaves pale blue with dark purple rachises.

Geographic Range: Florida, Mexico, Panama, Costa Rica, Mexico, Puerto Rico, Brazil.

Natural History: Found from 5-37 m depth.

Remarks: Originally described as a member of the genus *Chromodoris*, the radular morphology is more similar to that of members of the genus *Mexichromis*.

photos J. Hamann - left: animal with egg mass, La Parguera, Puerto Rico; right: gill, Cancún, Mexico

Family DORIDIDAE Rafinesque, 1815

Diagnosis: Body oval to elongate, covered by dorsal tubercles. Tubercles conical or rounded, simple, normally with no spicules protruding. Rhinophores and gill retractable; gill generally composed of unipinnate leaves. Ventral side with the anterior border of the foot transversely grooved but not longitudinally notched. Mouth area swollen with no true oral tentacles but conical lateral prolongations, Digestive system with a labial cuticle devoid of jaws. Radular teeth generally hook-shaped, simple or denticulate, in *Aldisa* extremely elongate. Reproductive system with an undivided prostate.

Biology: All members of the Dorididae feed on sponges and are normally found in rocky substrates. This group is particularly diverse in temperate and cold waters although numerous tropical species also occur.

Remarks: It is not clear whether this is a natural group or just an artificial subdivision of dorids lacking advanced features.

Doris verrucosa
Linnaeus, 1758

Size: Up to 70 mm long

photo © T. Gosliner - Tampa Bay, Florida

Doris bicolor
(Bergh, 1884)

Size: Up to 17 mm long

photos J. Hamann - Piton Bay, St. Lucia; left: light form; right: dark form

photos A. Valdés - SEM of *Doris kyolis*; left: inner radular teeth, scale bar = 30 µm; center: mid-lateral radular teeth, scale bar = 30 µm; right: outer radular teeth, scale bar = 30 µm

Doris verrucosa Linnaeus 1758: 653.
[synonyms from other areas not included]

Description: Background color yellowish gray to brown with darker shades forming two bands between the rhinophores and the gill. Dorsum covered with large, rounded tubercles of different sizes. Rhinophores and gill the same color as the dorsum and the sheaths surrounded by large tubercles. Gill composed of up to 8 unipinnate leaves.

Geographic Range: Amphiatlantic; Western Atlantic: Massachusetts, Virginia, South Carolina, Georgia, Florida, Costa Rica, Brazil.

Staurodoris bicolor Bergh 1884: 655-657, pl.69, figs. 2-7.

Description: Background color variable, from translucent yellowish gray to dark brown or black. Dorsum covered with rounded tubercles of different sizes, normally connected by a network of spiculose ridges. Rhinophores white; gill the same color as the dorsum. Rhinophoral and branchial sheaths surrounded by tubercles. Gill composed of up to 15 unipinnate leaves.

Geographic Range: Amphiatlantic; Western Atlantic: Florida, St. Lucia.

photo J. Hamann - Sewage Beach, Florida

Doris bovena
Er. Marcus, 1955

Size: Up to 17 mm long

photo J. Hamann - Miami, Florida

Doris kyolis
(Ev. Marcus & Er. Marcus, 1967)

Size: Up to 30 mm long

photo A. Valdés - Key Largo, Florida

Doris ilo
(Er. Marcus, 1955)

Size: Up to 25 mm long

Doris bovena Er. Marcus 1955: 131-134, fig. 109-118.

Description: Background color yellowish gray with irregular dark brown patches all over the dorsum. Dorsum covered with low, rounded tubercles. Rhinophores elongate, having the same color as the rest of the body. Gill dark brown with up to 12 leaves.

Geographic Range: Florida, Honduras, Venezuela, Aruba, Curaçao, Brazil.

Remarks: This species is distinguishable from *Doris kyolis* by having simple outer lateral teeth in the radula with no denticulations.

Siraius kyolis Ev. Marcus & Er. Marcus 1967a: 64-66, figs. 80-84.
Doris fretterae Thompson 1980: 84-86, figs. 7E, 8A-C.

Description: Background color from yellowish gray to greenish gray, blue or black, with some scattered opaque white spots. Dorsum with numerous small, conical to rounded tubercles. Rhinophores and gill with the same color as the rest of the body. Foot sole bright orange. Gill with up to 11 unipinnate branchial leaves.

Geographic Range: Florida, Venezuela, Puerto Rico, Virgin Islands, Bahamas, Curaçao, Barbados, St. Lucia, St. Vincent & the Grenadines, Grenada.

Natural History: Feeds on sponges; the color of the slug is similar to the color of the sponge on which it feeds.

photo A. Valdés - Curaçao.

Siraius ilo Er. Marcus 1955: 135-137, fig. 119-125.

Description: Background color uniformly brownish gray. Rhinophores and gill with the same color as the rest of the body. Dorsum covered by rounded tubercles. Gill with up to 22 unipinnate leaves.

Geographic Range: Florida, Trinidad & Tobago, Brazil.

Remarks: This species is recognizable by the large number of branchial leaves compared to other species.

Doris sp.

Size: Up to 10 mm long

photo A. Valdés - Curaçao

Aphelodoris antillensis
Bergh, 1879

Size: Up to 30 mm long

photos J. Hamann - left: Cantoy, Mexico; right: Cayman Islands

Aldisa sp.

Size: Up to 22 mm long

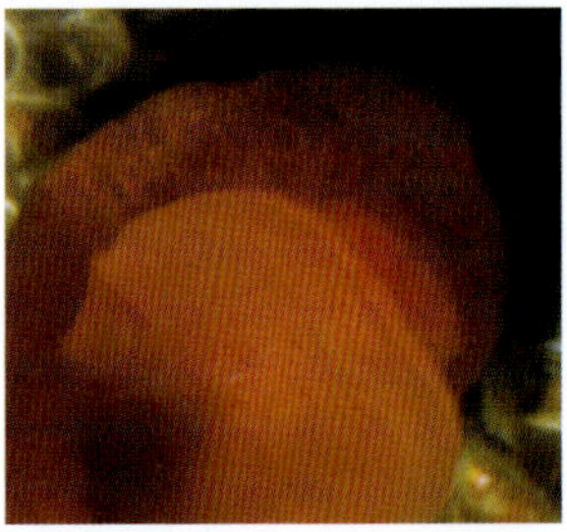

photos J. Hamann - left: adult specimen, St. Vincent & the Grenadines; right: head, St Lucia

Description: Background color uniformly yellowish gray. Dorsum covered by stalked, irregular green tubercles, larger around the visceral hump, interconnected by a complex network of spiculose ridges. Rhinophores and gill with the same color as the rest of the body.

Geographic Range: Curaçao.

Remarks: This species is probably undescribed; it is assigned to the genus *Doris* because it has a grooved but not notched anterior end of the foot, the dorsum covered by tubercles with no protruding spicules and a radula with hook-shaped teeth.

Aphelodoris antillensis Bergh 1879b: 108-113.
Doris bistellata Verrill 1900: 548, pl. 66, fig. 2.

photo J. Hamann - Aruba, head

Description: Background color translucent yellowish cream, with a very variable pattern of brown and opaque white pigmentation. Rhinophores and gill translucent cream with brown and white dots.

Geographic Range: Florida, Mexico, Costa Rica, Panama, Venezuela, Aruba, Curaçao, Cayman Islands, Jamaica, Dominican Republic, Puerto Rico, Virgin Islands, Antigua, St. Maarten/St. Martin, Guadeloupe, Martinique, Antigua, St. Lucia, St. Vincent & the Grenadines, Grenada, Bermuda, Barbados.

Natural History: When disturbed the animal swims by vigorous flexions of the body.

Description: Background color dark red, with a number of yellow dots on the dorsum. Rhinophores and gill uniformly red. Dorsum with two oval crater-like depressions on the midline.

Geographic Range: St. Vincent & the Grenadines, Puerto Rico, St. Lucia.

Remarks: This is probably an undescribed species and the first record of the genus *Aldisa* for the Caribbean Sea.

Family DISCODORIDIDAE Bergh, 1891

Body oval to elongate, covered by dorsal tubercles and/or dorsal ridges. Tubercles conical, rounded, or elongate, simple, with protruding spicules, or forming complex caryophyllidia with a ring of protruding spicules and cilia. Rhinophores and gill retractable; gill generally composed of multipinnate leaves. Ventral side with the anterior border of the foot transversely grooved and longitudinally notched. Mouth area small, with conical or tentacular oral tentacles. Digestive system with a labial cuticle with or without jaws. Radular teeth generally hook-shaped, simple or denticulate, in *Rostanga* extremely elongate. Reproductive system with a divided prostate; copulatory spines, penial spines and accessory glands present in some groups.

Biology: All members of the Discodorididae feed on sponges and are normally found on rocky substrates. Some species are able to autotomize sections of the mantle margin when disturbed.

Remarks: Several Caribbean species of Discodorididae have been described based on preserved specimens and are impossible to recognize; numerous species also remain undescribed.

photo A. Valdés - Puerto Vargas, Costa Rica

Discodoris evelinae
Er. Marcus, 1955

"Eveline's dorid"

Size: Up to 110 mm long

photo J. Hamann - left: light specimen, Martinique; right: dark specimen, Guadeloupe

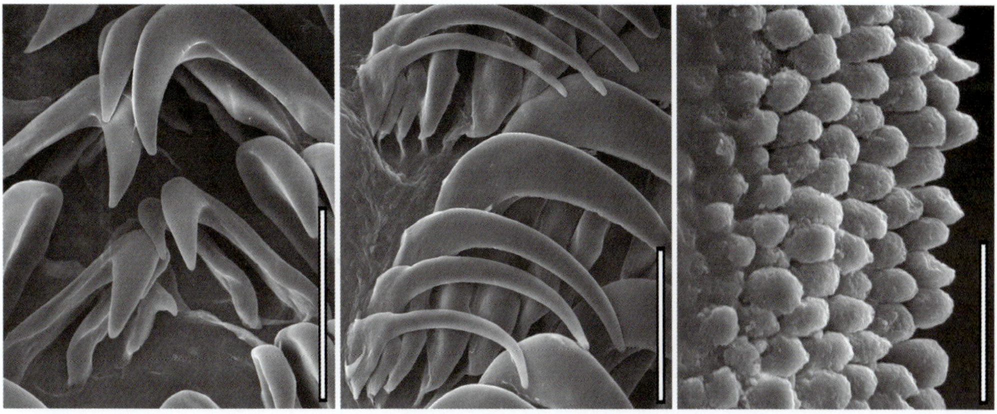

photos A. Valdés - SEM of *Diaulula hummelincki*; left: central radular teeth, scale bar = 50 µm; center: outer radular teeth, scale bar = 50 µm; right: jaw rodlets, scale bar = 20 µm

Discodoris evelinae Er. Marcus 1955: 143-147, fig. 141-150.
Discodoris hedgpethi Ev. Marcus & Er. Marcus 1960b: 254-256, figs. 7-11.

Description: Background color very variable, from cream or beige to purplish brown, with dark brown, black and opaque white patches and spots. Rhinophores and gill having the same general color as the dorsum, but with white tips. Dorsum covered by numerous conical tubercles. Underside greyish with numerous brown spots.

Geographic Range: Florida, Texas, Costa Rica, Honduras, Panama, Colombia, Venezuela, Bahamas, Cayman Islands, Puerto Rico, Jamaica, Barbados, Martinique, St. Lucia, Guadeloupe, St. Vincent & the Grenadines, Brazil.

Natural History: Animals hide under rocks and autotomize parts of the mantle margin when disturbed.

photo A. Valdés - rhinophore, Costa Rica

photos A. Valdés - Puerto Vargas, Costa Rica; left: anus and gill; right: ventral view

photo J. Hamann - Lake Surprise, Florida

Discodoris sp.

Size: Up to 50 mm long

photo © F. Burek - Exumas, Bahamas

Thordisa diuda
Er. Marcus, 1955

Size: Up to 36 mm long

photo J. Hamann - Cape Salomon, Martinique

Thordisa lurca
(Ev. Marcus & Er. Marcus, 1967)

Size: Up to 36 mm long

Description: Background color pale gray with numerous black patches and some opaque white dots. Body flattened, covered by conical tubercles. Rhinophores and gill having the same color as the dorsum with white tips.

Geographic Range: Florida.

Thordisa diuda Er. Marcus 1955: 140-143, figs. 133-140.

Description: Background color dirty yellow with a series of dark spots on the dorsum. Dorsum densely covered with long, conical papillae. Rhinophores and branchial leaves having the same color as the rest of the body.

Geographic Range: Jamaica, Bahamas, Brazil.

Nuvuca lurca Ev. Marcus & Er. Marcus 1967b: 621-623, figs. 48-50.

Description: Background color uniformly yellow. Dorsum covered with small, conical tubercles and several long, lighter colored tentacular papillae. Rhinophores pale brown with black tips. Branchial leaves yellowish with white tips.

Geographic Range: Costa Rica, Martinique, Brazil.

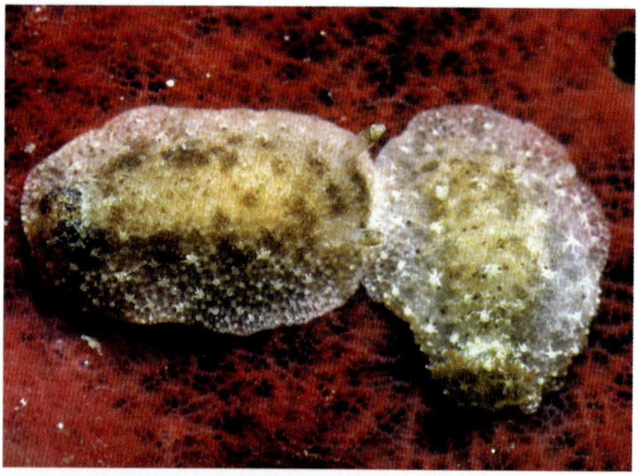

Geitodoris cf. *planata*
(Alder & Hancock, 1846)

Size: Up to 65 mm long

photo J. Hamann - Beaumont Point, St. Lucia

Geitodoris immunda
Bergh, 1894

Size: Up to 43 mm long

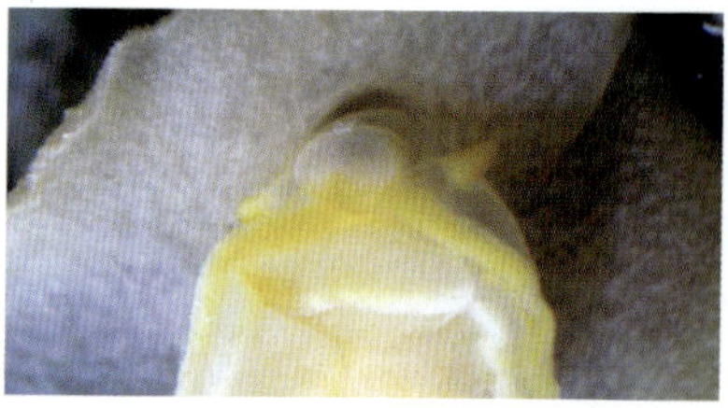

photo J. Hamann - Miami, Florida; left: adult specimen; right: ventral view

Geitodoris pusae
(Er. Marcus, 1955)

Size: Up to 47 mm long

photo J. Hamann - Discovery Bay, Jamaica; left: adult specimen; right: egg mass

Doris planata Alder & Hancock 1846: 292-293.
Doris complanata Verrill 1880: 399.

Description: Background color greyish brown with some dark brown irregular patches. The color fades out towards the mantle margin. Dorsum covered by rounded, stalked tubercles. Larger tubercles with opaque white pigment around them. Rhinophores and gill having the same color as the rest of the body with white tips.

Geographic Range: Amphiatlantic; Western Atlantic: New Jersey, St. Lucia.

Geitodoris immunda Bergh 1894: 167-168, pl. 5, figs. 6-12.

Description: Background color greyish brown with numerous opaque white dots and some dark brown areas. Dorsum with a complex network of low ridges covering the entire surface, with some conical tubercles scattered all over. Branchial sheath with a characteristic wavy edge. Rhinophores and branchial leaves brown with white tips.

Geographic Range: Costa Rica, Venezuela, Brazil.

Discodoris pusae Er. Marcus 1955: 147-151, fig. 151-165.

Description: Background color dirty yellow with some scattered, irregular opaque white patches. Dorsum with a complex network of low ridges covering the entire surface, with some large, elongate tubercles scattered all over. Rhinophores and branchial leaves having the same color as the dorsum with opaque white and brown spots.

Geographic Range: Costa Rica, Jamaica, Martinique, Brazil.

Paradoris mulciber
(Ev. Marcus, 1971)

Size: Up to 40 mm long

photos J. Hamann - Fort Vieux Bay, St. Lucia; left: dorsal view; right: oral tentacles

Paradoris sp.

Size: Up to 50 mm long

photo A. Valdés - Bocas del Toro, Panama

Hoplodoris sp.

Size: Up to 11 mm long

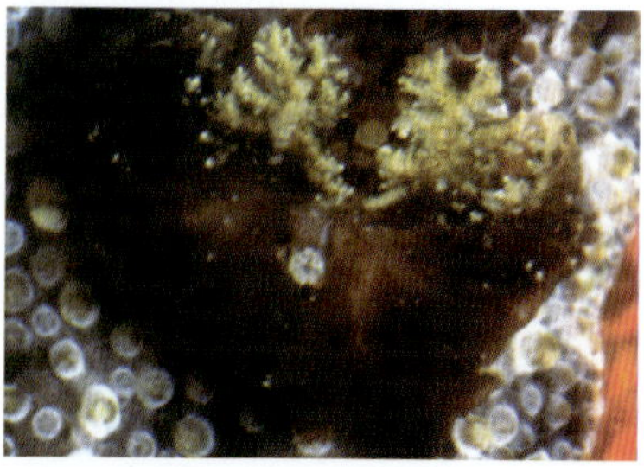

photos J. Hamann - left: Cayman Islands; right: gill, Pt. Antonio, Jamaica

Percunas mulciber Ev. Marcus 1971: 945-947, figs. 45-49.

Description: Background color uniformly cream to pale gray with a few black spots. Dorsum with low, rounded tubercles, somewhat darker than the body in some specimens. Rhinophores and gill pale ochre with white spots and white tips.

Geographic Range: Costa Rica, Panama, St. Maarten/St. Martin, St. Lucia, Brazil.

photo J. Hamann - St. Maarten/St. Martin

Description: Background color uniformly dark brown with a few white dots. Dorsum with several large, conical tubercles, somewhat lighter than the body. Rhinophores and gill dark brown with paler tips.

Geographic Range: Panama.

Remarks: This appears to be an undescribed species.

Description: Background color reddish gray. Dorsum covered by large, single, rounded tubercles. Opaque white rings at the base of the largest tubercles, especially in those situated in two rows between the rhinophores and the gill. Rhinophores having the same color as the dorsum with white tips. Branchial leaves short, pointing forward in the living animal; dark brown, with the two anterior leaves much longer than the rest, having creamy apices.

Geographic Range: Cayman Islands, Jamaica.

Remarks: This is probably an undescribed species and the first record of the genus *Hoplodoris* for the Atlantic Ocean.

Platydoris angustipes
(Mörch, 1863)

Size: Up to 150 mm long

photo J. Hamann - French Key, Turks & Caicos

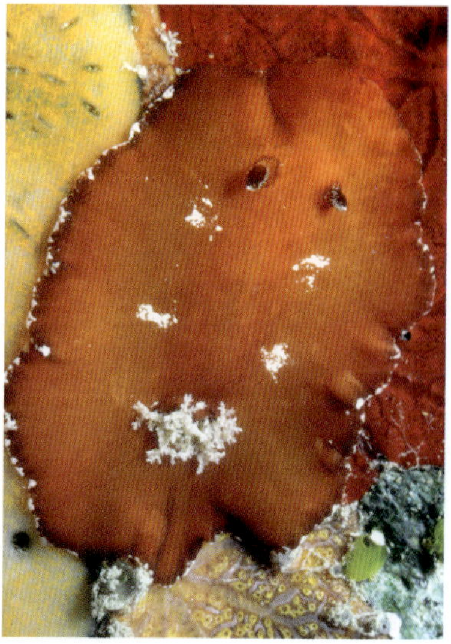

photos J. Hamann - left: large specimen, Aruba; right: Guanaja, Honduras

Rostanga byga
Er. Marcus, 1958

Size: Up to 20 mm long

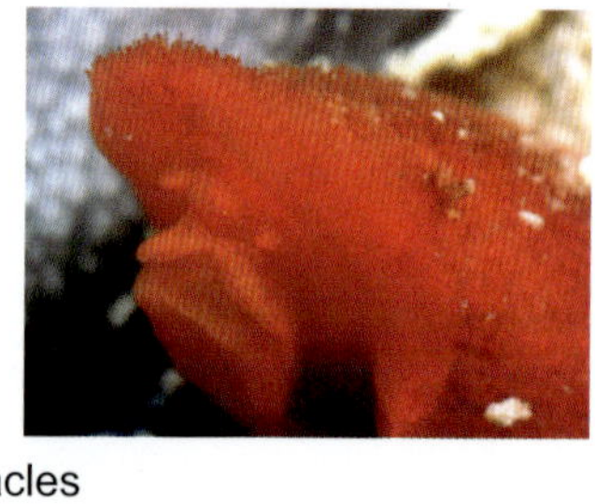

photos J. Hamann - St. Lucia; left: dorsal view; right oral tentacles

Doris (Argus) angustipes Mörch 1863: 32.
Platydoris angustipes var. alaleta Bergh 1877: 505, pl. 58, figs. 13-18.
Platydoris rubra White 1952: 118.

Description: Background color ranges from reddish brown, to red or orange. Scattered white specks, normally clustered in 3-4 dense groups especially on the mantle margin. Mantle edge with a thin translucent line ranging in color from straw to dark brown, sometimes with scattered white triangular markings. Dorsum covered with sessile caryophyllidia. Rhinophores dark brown. Branchial leaves translucent straw-colored with opaque white specks.

Geographic Range: Florida, Mexico, Honduras, Costa Rica, Panama, Cayman Islands, Dry Tortugas, Cuba, Jamaica, Aruba, Puerto Rico, Virgin Islands, St. Maarten/St. Martin, St. Barthelemy, Turks & Caicos, Antigua, St. Lucia, Martinique, Barbuda, St. Vincent & the Grenadines, Grenada, Trinidad & Tobago, Brazil.

Natural History: Inhabits reefs. Feeds on sponges.

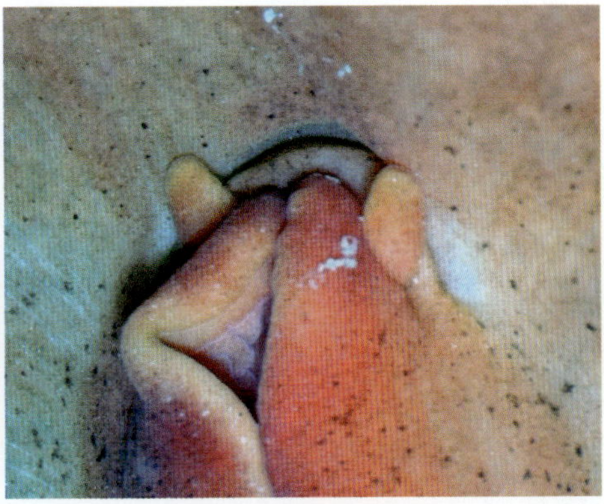

photos A. Valdés - Manzanillo, Costa Rica; left: dorsal view; right: oral tentacles

---

Rostanga byga Er. Marcus 1958: 22-25, figs. 34-36.

Description: Background color uniform red with dark spots in some specimens. Dorsum completely covered by densely arranged, large caryophyllidia. Rhinophores red, with a few vertical lamellae. Gill composed of unipinnate, normally dark red leaves.

Geographic Range: St. Lucia, Mexico, Bahamas, Bermuda, Brazil, Argentina.

Taringa telopia
Er. Marcus, 1955

Size: Up to 40 mm long

photos J. Hamann - left: light specimen, Aruba; right: dark specimen, Piton Bay, St. Lucia

photo J. Hamann - French Key, Turks & Caicos

Jorunna spazzola
(Er. Marcus, 1955)

Size: Up to 18 mm long

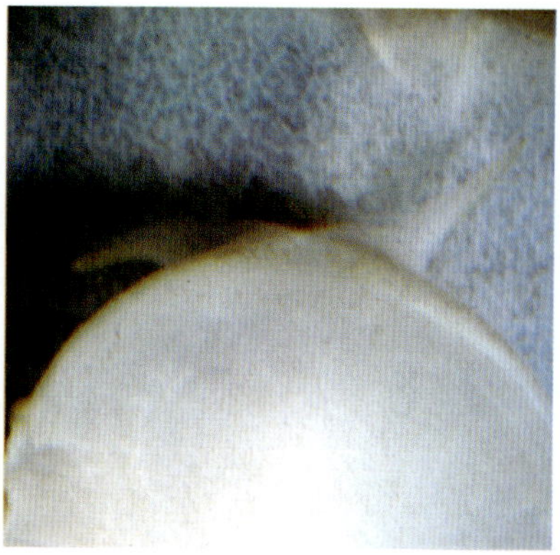

photos J. Hamann - Roatán, Honduras; left: adult specimen; right: oral tentacles

Taringa telopia Er. Marcus 1955: 152-155, figs. 166-179.
Taringa telopia disa Ev. Marcus & Er. Marcus 1967a: 87-88, figs. 113-114.

Description: Background color variable from dirty yellow to dark brown, almost black, with numerous irregular opaque white flecks and larger dark brown spots. Rhinophoral and branchial sheaths surrounded by opaque white radiating spots. Dorsum covered with small caryophyllidia. Gill always pale gray or cream. Rhinophores also pale with dark spots and white apices.

Geographic Range: Dominican Republic, Aruba, St. Lucia, Barbados, Brazil.

Awuka spazzola Er. Marcus 1955: 156-158, fig. 180-192.
?*Discodoris mortenseni* Ev. Marcus & Er. Marcus 1963: 30-32, figs. 36-39.

Description: Background color translucent gray with a few, somewhat darker patches over the entire dorsum, larger around the visceral hump. Dorsum covered with small caryophyllidia. Mantle margin surrounded by small opaque white glands. Rhinophores and gill the same color as the rest of the body. Branchial leaves very short.

Geographic Range: Florida, Honduras, Cuba, Curaçao, Barbados, Virgin Islands, Turks & Caicos, Brazil.

Natural History: This species is able to change color in minutes responding to environmental factors. When an animal is found in the field is normally pure white, becoming gray after being put in a container.

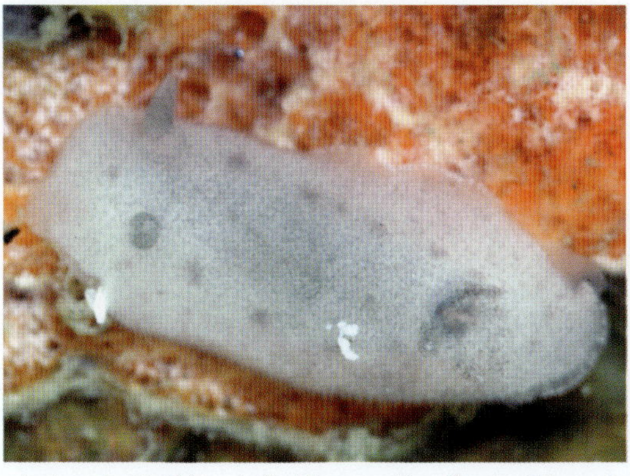

photos A. Valdés - left: Manzanillo, Costa Rica; right: Key Largo, Florida

Sclerodoris prea
(Ev. Marcus & Er. Marcus, 1967)

Size: Up to 40 mm long

photo J. Hamann - Sewage Beach, Florida

Sclerodoris worki
(Ev. Marcus & Er. Marcus, 1967)

"Work's dorid"

Size: Up to 50 mm long

photo J. Hamann - Cayman Islands

photos A. Valdés - Curaçao; left: dorsal view; right: ventral view

Anisodoris prea Ev. Marcus & Er. Marcus 1967a: 70-72, figs. 90-93.

Description: Background color cream-brown with numerous dark brown spots. Dorsum covered with numerous tubercles forming cauliflower-like structures with an apical knob surrounded by many smaller tubercles. Larger tubercles arranged in two rows from the rhinophores to the gill, leaving a longitudinal depression in the center, with black patches free of tubercles. Similar but smaller areas occur all over the dorsum. Rhinophores very elongate, cream with dark spots. Gill composed of multipinnate branchial leaves, pale gray with opaque white spots.

Geographic Range: Florida, Bahamas, Jamaica, Barbados.

Anisodoris worki Ev. Marcus & Er. Marcus 1967a: 66-70, figs. 85-89.

Description: Background color variable from cream-brown to dark gray with some black patches. Dorsum covered with numerous tubercles forming cauliflower-like structures with an apical knob surrounded by many smaller tubercles. Larger tubercles orange, sometimes with white pigment around the bases. Rhinophores very elongate, cream to dark brown with white dots. Gill composed of multipinnate branchial leaves, with the same color as the dorsum.

Geographic Range: Florida, Honduras, Cayman Islands, Curaçao, St. Lucia.

photos J. Hamann - left: Piton Bay, St. Lucia; right: Roatán, Honduras

photo J. Hamann - Miami, Florida

Diaulula greeleyi
(MacFarland, 1909)

Size: Up to 47 mm long

photo J. Hamann - Roatán, Honduras

Diaulula phoca
(Ev. Marcus & Er. Marcus, 1963)

Size: Up to 50 mm long

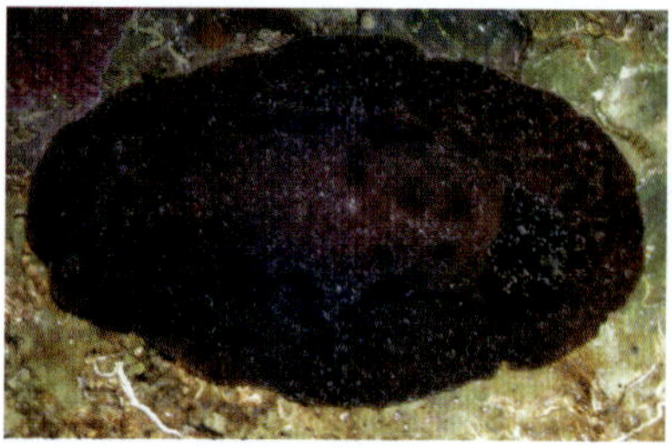

photos A. Valdés - Panama

photos A. Valdés - Key Largo, Florida

Diaulula farmersi
Valdés, 2004

Size: Up to 16 mm long

Peltodoris greeleyi MacFarland 1909: 84-88, pl. 15, figs. 77-82.
Peltodoris nayarita Ortea & Llera 1981: 47–51, figs. 1-4.

Description: Background color dirty yellow with numerous dark brown patches. Dorsum covered with small caryophyllidia. Rhinophoral and branchial sheaths very elevated. Rhinophores black. Branchial leaves numerous, dark gray with white apices. Mantle margin with a few opaque white glands.

Geographic Range: Eastern Pacific, Western Atlantic: Florida, Bahamas, Brazil.

Discodoris phoca Ev. Marcus & Er. Marcus 1967a: 78-80, figs. 99-101.

Description: Background color dark purplish brown with numerous white dots. Dorsum covered with small caryophyllidia. Rhinophores and gill having the same color as the rest of the dorsum.

Geographic Range: Florida, Honduras, Costa Rica, Panama.

Diaulula farmersi Valdés 2004: 6-7, figs. 1B, 4-5.

Description: Background color dirty yellow, some specimens with light brown patches. Dorsum covered with small caryophyllidia. Rhinophoral and branchial sheaths low. Rhinophores black. Branchial leaves numerous, white to pale gray. Mantle margin with numerous opaque white glands.

Geographic Range: Florida, Puerto Rico, Virgin Islands, Bahamas.

Diaulula hummelincki
(Ev. Marcus & Er. Marcus, 1963)

"Hummelinck's dorid"

Size: Up to 50 mm long

photos J. Hamann & A. Valdés - left: Groupers, St. Maarten (JH); right: ventral view, Curaçao (AV)

Diaulula alba
(White, 1952)

Size: Up to 32 mm long

photo J. Hamann - Peter Island, Virgin Islands

Atagema browni
Thompson, 1980

Size: Up to 24 mm long

photo © A. Bernecker - Puerto Vargas, Costa Rica

Peltodoris hummelincki Ev. Marcus & Er. Marcus 1963: 27-30, figs. 32-35.

Description: Background color translucent gray with several large black or gray spots broken by a network of whitish lines. Dorsum covered with small caryophyllidia, some of them pigmented with white. Rhinophoral and branchial sheaths with small, pointed white papillae. Rhinophores and gill pale brown with white tips and some white spots. Mantle margin edged by small white papillae.

Geographic Range: Aruba, Curaçao.

Discodoris alba White 1952: 113, fig. 12, pl. 6, fig. 5.

Description: Background color translucent white with some opaque white dots. Dorsum covered with small caryophyllidia. Rhinophores and gill having the same color as the rest of the body.

Geographic Range: Dry Tortugas, Virgin Islands.

Remarks: The radular morphology of the specimen photographed matches the original description of *Discodoris alba*, which is transferred to *Diaulula* because of the presence of caryophyllidia and other anatomical features.

Atagema browni Thompson 1980: 87, figs. 7F, 9B-C.

Description: Background color pale gray with large black patches. Entire dorsum covered by large caryophyllidia forming elevated aggregations on the center of the dorsum. Rhinophoral sheaths also elevated. Branchial sheath composed of three large valves that close when the gill is retracted. Gill and rhinophores the same color as the dorsum.

Geographic Range: Costa Rica, Cayman Islands, Jamaica.

Discodorididae sp. 1

Size: Up to 20 mm long

photos J. Hamann - Aruba; left: dorsal view of adult specimen; right: egg mass

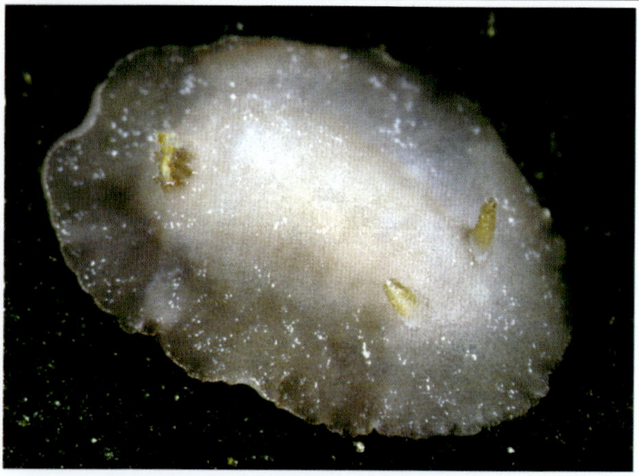

Discodorididae sp. 2

Size: Up to 33 mm long

photo J. Hamann - Crocus Bay, Anguilla

Discodorididae sp. 3

Size: Up to 76 mm long

photo J. Hamann - Roatán, Honduras

Description: Background color translucent white with some opaque white dots and large pale brown patches. Dorsum covered with small caryophyllidia. Rhinophores and gill having the same color as the rest of the body.

Geographic Range: Aruba.

Remarks: This appears to be an undescribed species.

Description: Background color translucent white with some opaque white dots. Dorsum covered with small caryophyllidia. Rhinophores and gill pale brown with white spots.

Geographic Range: Anguilla.

Remarks: This appears to be an undescribed species.

Description: Background color translucent white with some opaque white dots and shades of brown and gray. Dorsum covered with small caryophyllidia. Rhinophores pale brown. Gill having the same color as the rest of the body with white tips.

Geographic Range: Honduras.

Remarks: This appears to be an undescribed species.

photo J. Hamann - Samaná, Dominican Republic

Discodorididae sp. 4

Size: Up to 55 mm long

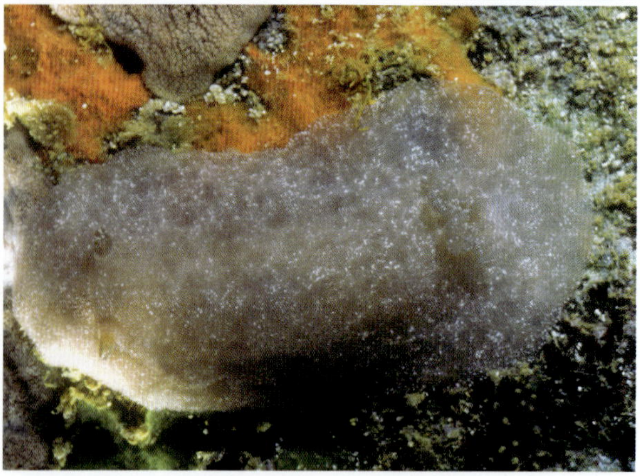

photo J. Hamann - Samaná, Dominican Republic

Discodorididae sp. 5

Size: Up to 30 mm long

photo J. Hamann - Belize

Discodorididae sp. 6

Size: Up to 9 mm long

Description: Background color translucent gray with opaque white dots and pale brown patches. Dorsum covered with small caryophyllidia. Rhinophores dark brown. Gill having the same color as the rest of the body with white tips.

Geographic Range: Dominican Republic.

Remarks: This appears to be an undescribed species.

Description: Background color translucent gray with numerous opaque white dots all over the dorsum. Dorsum covered with small caryophyllidia. Rhinophores pale gray with white spots. Gill with the same color as the rest of the body.

Geographic Range: Dominican Republic.

Remarks: This appears to be an undescribed species.

Description: Background color pale brown with numerous dark brown spots and opaque white dots all over the dorsum. Rhinophores and gill pale gray, with no distinctive markings.

Geographic Range: Belize.

Remarks: This appears to be an undescribed species.

Family DENDRODORIDIDAE O'Donoghue, 1924

Diagnosis: Dorsum smooth or covered with tubercles, in *Doriopsilla* stiffened by spicules. Head with oral tentacles reduced and fused together, giving the appearance that there are no oral tentacles. Anterior border of the foot simple or notched. Gill dorsal, composed of a circlet of branchial leaves completed by the anus, on the right side in *Doriopsilla* and posteriorly in *Dendrodoris*.

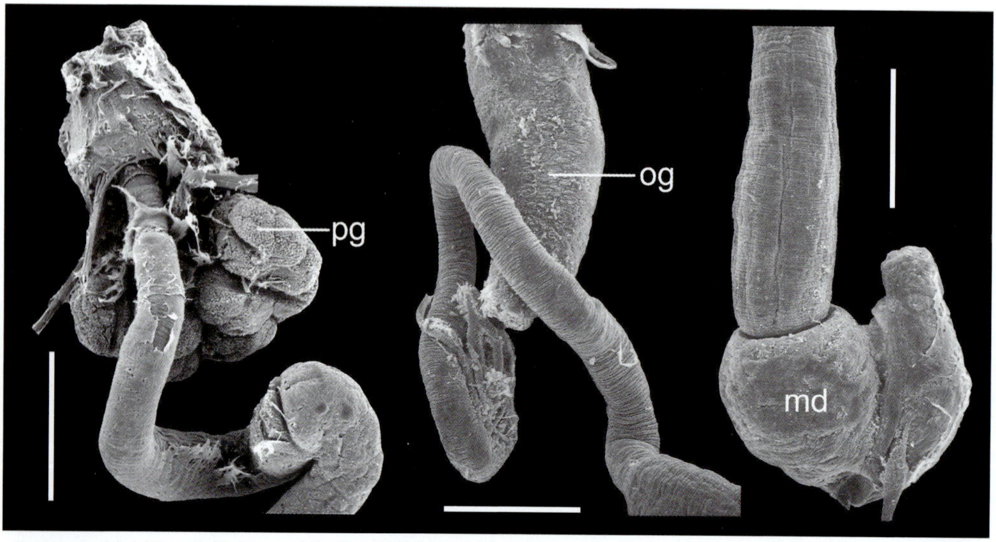

photos A. Valdés - left: buccal bulb of *Dendrodoris* showing the ptyaline glands (pg), scale bar = 1.5 mm; center, buccal bulb of *Doriopsilla* showing the oral glands (og), scale bar = 250 μm; right: esophagus of *Doriopsilla* showing the muscular dilation (md), scale bar = 300 μm.

Dendrodoris warta

Marcus & Gallagher, 1976

"Warty dendrodorid"

Size: Up to 120 mm long

photos © T. Gosliner & © C. Cox - left: Sarasota Bay, Florida (TG); right: Florida (CC)

photos A. Valdés - left: gill of *Dendrodoris* showing the anus (a) posterior to the branchial leaves, scale bar = 1.5 mm; right: gill of *Doriopsilla* showing the anus (a) on the right of the circlet of branchial leaves, scale bar = 1.5 mm.

Radula and labial cuticle completely absent, replaced by a series of oral glands on the buccal bulb in *Doriopsilla*, or ptyaline glands connected to the buccal bulb by a tube in *Dendrodoris*. Esophagus with a muscular or glandular dilation. Penis armed with hooks characteristic of each species. Vestibular or accessory glands present in a few species.

Biology: Members of this family are suctorial feeders on sponges. Many species have a characteristic spicy taste derived from drimane sesquiterpenoids that the nudibranchs synthesize independently from their food source and store in their mantle for defense. Dendrodorids are found in tropical, subtropical, and temperate regions of the Atlantic, Pacific and Indian Oceans.

Dendrodoris warta Marcus & Gallagher 1976: 353-356, figs. 1-9.

Description: Background color dark brown. Dorsum covered with large, irregular, pinkish beige wart-like tubercles. Gray rhinophores and brown gill.

Geographic Range: South Carolina, Florida, Venezuela.

Natural History: Egg mass composed of numerous large eggs (about 120 µm in diameter).

Remarks: This is a very uncommon species probably endemic to southeastern North America.

Dendrodoris krebsii
(Mörch, 1863)

"Krebs' dendrodorid"

Size: Up to 70 mm long

photo J. Hamann - Ranguana Cay, Belize

photos J. Hamann - left: Guanaja, Honduras; right: Florida

photos J. Hamann - left: Roatán, Honduras; right: Guanaja, Honduras

Doris krebsii Mörch 1863: 34-35
Doridopsis subpellucida Abraham 1877: 265-266, pl. 30, fig. 36.
Doridopsis atropos Bergh 1879a: 49-64.
Doriopsis krebsi pallida Bergh 1879a: 44-49.

Description: Background color highly variable, ranging from white to black, orange or light green, with or without spots of red, black, gray, or white. Vein-like markings on the mantle margin. Rhinophores and gill normally the same color as the rest of the body with white apices.

Geographic Range: Georgia, Florida, Mexico, Belize, Honduras, Costa Rica, Panama, Colombia, Venezuela, Aruba, Curaçao, Bahamas, Cuba, Dominican Republic, Cayman Islands, Jamaica, Virgin Islands, St. Maarten/St. Martin, Guadeloupe, Martinique, Barbados, Antigua, St. Lucia, St. Vincent & the Grenadines, Grenada, Brazil.

Natural History: Feeds on sponges. Hides under coral and rocks.

Remarks: This is one of the most common species of nudibranchs in the Caribbean. Similarly colored animals from the eastern Pacific belong to a different species.

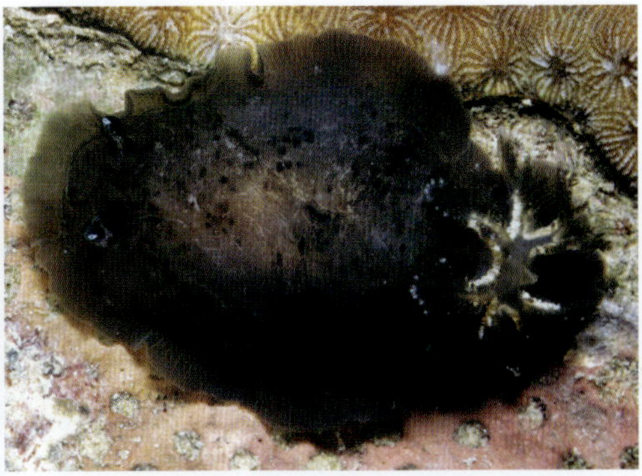

photos J. Hamann & A. Valdés - left: Aruba (JH); right: Bocas del Toro, Panama (AV)

photos J. Hamann - Aruba

Dendrodoris magagnai
Ortea & Espinosa, 2001

Size: Up to 26 mm long

photo © L. Moro Abad

Doriopsilla nigrolineata
Meyer, 1977

Size: Up to 30 mm long

photos J. Hamann & © H. Bertsch - left: Guanaja, Honduras (JH); right: egg mass, Panama

Doriopsilla espinosai
Valdés & Ortea, 1998

Size: Up to 15 mm long

photo J. Hamann - Bahamas

Dendrodoris magagnai Ortea & Espinosa 2001c: 52-53, fig. 13A-C, pl. 3, fig. E.

Description: Background color white with yellow or orange spots and dark areas composed of aggregations of black chromatophores. Mantle margin edged by a yellow line. Rhinophores and gill dark with white apices.

Geographic Range: Costa Rica.

Natural History: Found at 20 m depth.

Doriopsilla nigrolineata Meyer 1977: 304-305, fig. 3.

Description: Background color translucent white to orange, with a series of irregular black lines all over the dorsum. Dorsum covered with rounded tubercles with the bases densely speckled with white. White spots dispersed over the rest of the dorsum. Rhinophores and gill yellow.

Geographic Range: Panama, Honduras.

Natural History: Found at 3-6 m depth.

Remarks: Previously considered as a subspecies of *Doriopsilla areolata* by Valdés & Ortea (1997), examination of living specimens confirmed that it is a distinct species.

Doriopsilla espinosai Valdés & Ortea 1998: 389-3394, figs. 1-4.

Description: Background color translucent white to yellowish, with several large opaque white patches and red spots on the dorsum. Rhinophores and gill having a similar color to the rest of the body. Dorsum covered with small, conical tubercles.

Geographic Range: Cuba, Bahamas.

Natural History: Found at 1 m depth, under rocks. The egg mass consists of small ribbons with large eggs, suggesting this species may display direct development.

photo J. Hamann - egg masses, Bahamas

photo J. Hamann - Cayo Lobos, Puerto Rico

Doriopsilla* cf. *areolata
Bergh, 1880

Size: Up to 30 mm long

photos J. Hamann - Terre d'en Haut, Guadeloupe - left: dorsal view; right: ventral view

---

photo © T. Gosliner - Sarasota Bay, Florida

Doriopsilla pharpa
Er. Marcus, 1961

Size: Up to 18 mm long

Doriopsilla areolata Bergh 1880a: 318-326, pl. 11, figs. 3-11.

Description: Background color translucent yellow to orange. Dorsum covered with rounded tubercles. Opaque white circles surround some of the tubercles. Rhinophores and gill having the same color as the rest of the body.

Geographic Range: Eastern Atlantic; Western Atlantic: Virgin Islands, Puerto Rico, Martinique.

Remarks: This species was originally described from the Mediterranean. Valdés & Ortea (1997) synonymized several names and proposed a new classification including three subspecies. Caribbean records of *D. areolata* by Ev. Marcus & Er. Marcus (1967a) were regarded as uncertain. Caribbean specimens differ from most eastern Atlantic populations by lacking a network of white lines, but they are anatomically indistinguishable.

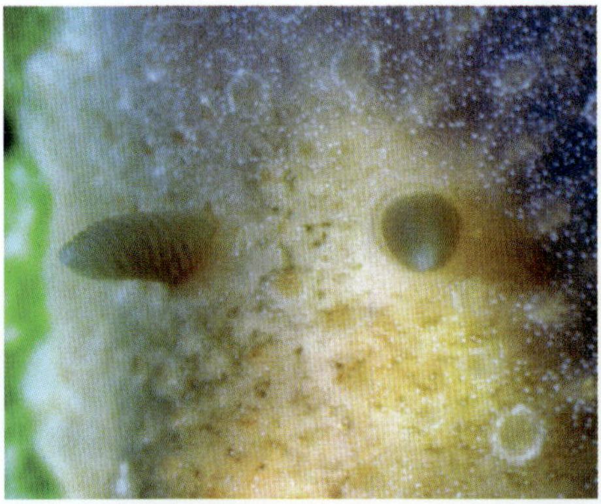
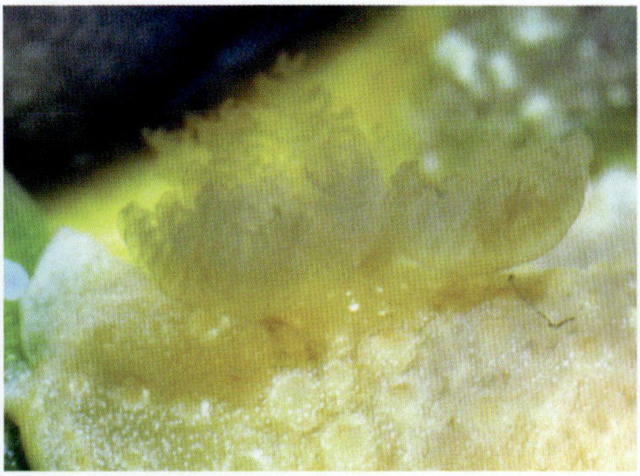

photos J. Hamann - Terre d'en Haut, Guadeloupe - left: rhinophores; right: gill

---

Doriopsilla pharpa Er. Marcus 1961: 146, figs. 19-21.
Doriopsilla leia Er. Marcus 1961: 144-146, figs. 15-18.

Description: Background color uniformly yellow to orange with numerous brown spots. Dorsum covered with small rounded tubercles. Rhinophores and gill having the same color as the rest of the body.

Geographic Range: Maryland, North Carolina, South Carolina, Georgia, Florida, Cuba, Virgin Islands.

Natural History: Feeds on burrowing sponges of the genus *Cliona*.

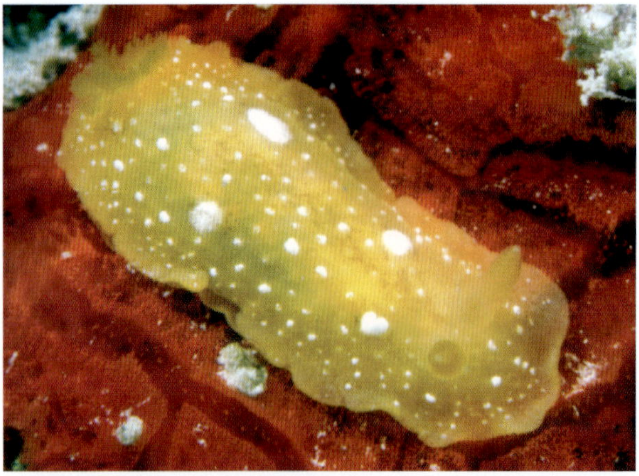

Doriopsilla sp. 1

Size: Up to 20 mm long

photo J. Hamann - left: yellow form, Aruba; right: orange form, Martinique

Doriopsilla sp. 2

Size: Up to 18 mm long

photo J. Hamann - Cayo Luis Peña, Puerto Rico

Doriopsilla sp. 3

Size: Up to 24 mm long

photo J. Hamann - left: Guanaja, Honduras; right: Roatán, Honduras

Description: Background color uniformly yellowish to orange with numerous white spots. Four larger white blotches occur on the center of the dorsum. Dorsum covered with small rounded tubercles. Rhinophores and gill with the same color as the rest of the body.

Geographic Range: Aruba, St. Vincent & the Grenadines, Grenada, Martinique.

Natural History: Found at 1-20 m depth.

Remarks: This is probably an undescribed species.

photo J. Hamann - orange form, St. Vincent & the Grenadines

Description: Background color uniformly translucent gray with small white spots. Network of spicules and viscera visible through the skin. Dorsum covered with small conical tubercles. Rhinophores and gill having the same color as the rest of the body.

Geographic Range: Puerto Rico.

Natural History: Found at 20 m depth.

Remarks: This is probably an undescribed species.

Description: Background color translucent white with a network of black lines. Dorsum covered with large, rounded, opaque white tubercles, larger near the center of the body. Rhinophores and gill opaque yellow.

Geographic Range: Honduras.

Natural History: Found at 3 m depth.

Remarks: This species is similar to *Doriopsilla nigrolineata* but it lacks the opaque white rings around the tubercles and the background color is more translucent. It is probably an undescribed species.

Family PHYLLIDIIDAE Rafinesque, 1814

Diagnosis: Dorsum covered with tubercles, normally stiffened by spicules. Head with oral tentacles reduced and often fused together. Anterior border of the foot simple. Dorsal gill absent, replaced by a series of ventro-lateral gills situated between the foot and the ventral end of the mantle margin. Radula and jaws completely absent, replaced by a series of oral glands situated on the heavily modified buccal bulb. Esophagus with a muscular expansion. Penis armed with hooks. Vestibular or accessory glands normally absent.

Biology: Members of this family are suctorial feeders on sponges. Many species have a characteristic odor derived from sesquiterpene isocyanides that the nudibranchs obtain from their prey and store in their mantle for defense. Phyllidiid chemicals are extremely toxic and can kill other organisms kept in aquaria together with the nudibranchs. Phyllidiids are exclusively found

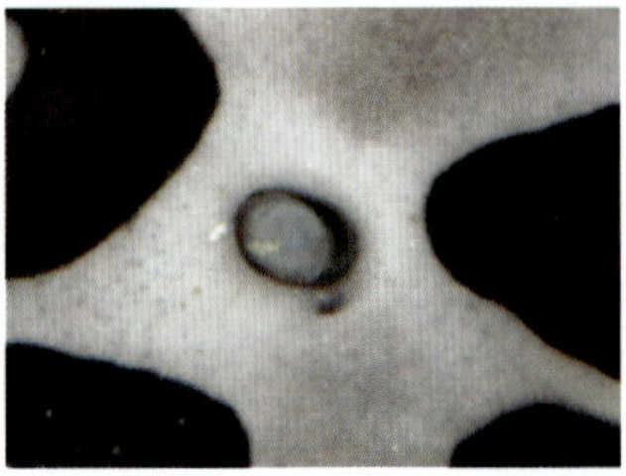

photo A. Valdés - anus of *Ceratophyllidia papilligera*.

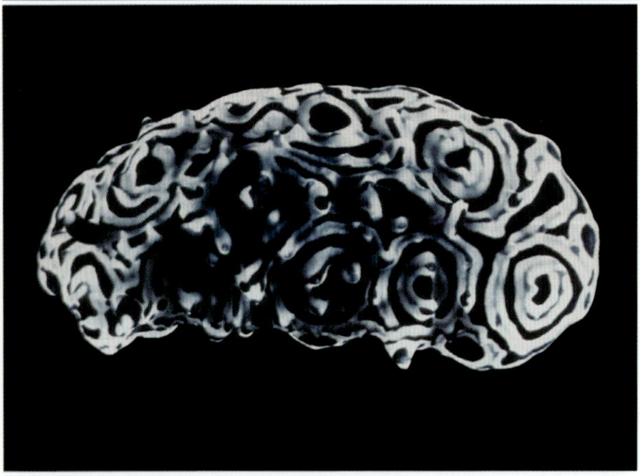

Phyllidiella molaensis
(Meyer, 1977)

Size: Up to 45 mm long

photo © K. Meyer - Galeta Point, Panama

Ceratophyllidia papilligera
(Bergh, 1890)

Size: Up to 30 mm long

photos J. Hamann & A. Valdés - left: Guadeloupe (JH); right: St. Thomas, Virgin Islands (AV)

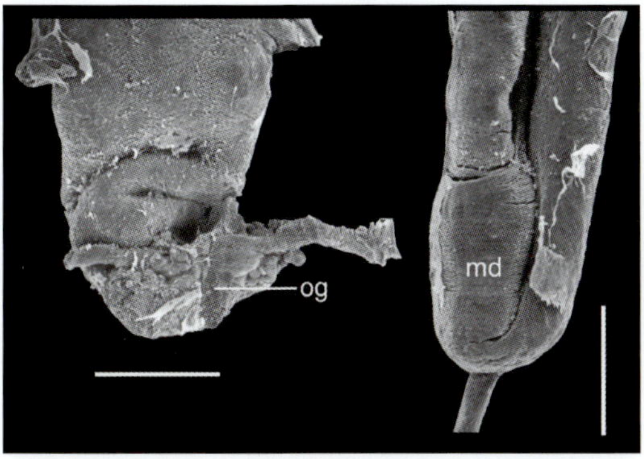

in tropical and subtropical regions of the Atlantic, Pacific and Indian Oceans, reaching areas below 700 m depth.

photos A. Valdés & J. Hamann - left: SEM photos of *Ceratophyllidia papilligera*; Left: buccal bulb showing oral glands (og) on the posterior end, scale bar = 1 mm; Right: esophagus showing the muscular expansion (md), scale bar = 600 μm (AV); right: Guadeloupe - ventro-lateral gill of *Ceratophyllidia papilligera* (JH)

Phyllidiopsis molaensis Meyer 1977: 305-306, fig. 4.

Description: Background color white with several sets of black and white concentric rings along the sides of the dorsum. Mantle covered with conical tubercles. Rhinophores black with white tips.

Geographic Range: Costa Rica, Panama.

Natural History: Lives on rocky substrates.

Remarks: Very rare species, seems to be endemic to the southwestern Caribbean.

Phyllidiopsis papilligera Bergh 1890: 176-178, pl. 2, figs. 7-14.

Description: Background color white with grayish black spots of varying size. Mantle covered with stalked tubercles. Rhinophores black with white apices.

Geographic Range: Florida, Mexico, Honduras, Costa Rica, Bahamas, Cuba, Cayman Islands, Virgin Islands, Guadeloupe.

Natural History: Found on rocky areas from 2-185 m depth.

photo J. Hamann - Guadeloupe, anterior end of the body

Family ARMINIDAE Iredale & O'Donoghue, 1923

Diagnosis: Body elongate, with a distinct oral veil and short, lateral tentaculate lobes. Dorsum often with longitudinal ridges. Mantle and foot approximately equal in length and connected posteriorly. A transverse flap, lobe or ridge (caruncle) present in some species immediately anterior to the rhinophores. Rhinophores with series of vertical lamellae, subdivided into secondary and tertiary plates. In *Armina*, ventral surface of the mantle margin with leaf like gills, and a complex system of oblique folds (secondary lamellae). Digestive system with two large and thick jaws and a broad radula with a triangular central tooth and numerous denticulate lateral teeth.

Biology: Species of Arminidae are commonly found on soft sandy bottoms, from shallow waters down to great depths. They feed on sea pens using their jaws and radula to bite off pieces of flesh.

Remarks: There are four species of *Armina* found in the tropical western Atlantic, but only two of them are included in this book. The deep-sea species *Armina juliana* Ardila & Díaz, 2002 has only

photo © K. Clark - Florida

Armina muelleri
(Ihering, 1886)

Size: Up to 32 mm long

photo J. Hamann - Chimana Grande, Venezuela

Armina wattla
Ev. Marcus & Er. Marcus, 1967

Size: Up to 12 mm long

been found in Colombia below 460 m depth. *Armina elongata* Ardila & Valdés, 2004, also from Colombia is a shallow water species, but no living specimens have ever been found.

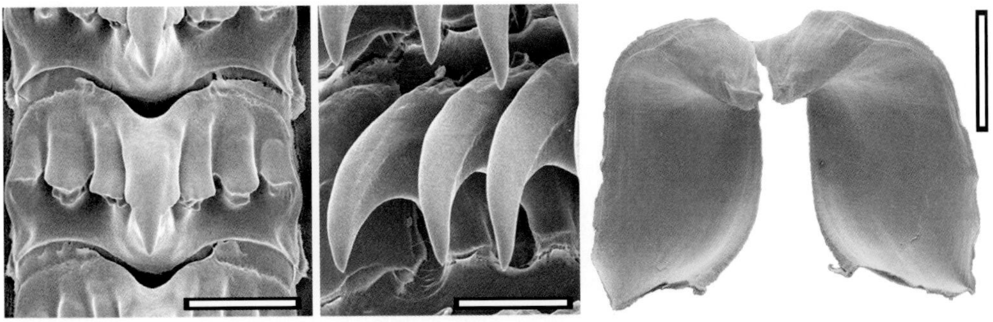

photos © N. Ardila - SEM of *Armina muelleri*; left: central radular teeth, scale bar = 60 µm; center: mid-lateral radular teeth, scale bar = 60 µm; right: jaws, scale bar = 1 mm

Pleurophyllidia mülleri Ihering 1886: 223-228, pl. 9, fig. 1.

Description: Background color variable from red to black, with a series of longitudinal yellow ridges on the dorsum, situated close to each other. Velum red or black with a yellow rim. Foot wider than the mantle. Rhinophores yellow, small, covered by the mantle in the living animal.

Geographic Range: Colombia, Venezuela, Dominica, Brazil.

Natural History: Probably feeds on the sea pens *Renilla reniformis* and *Renilla muelleri.*

Armina wattla Ev. Marcus & Er. Marcus 1967c: 213-216, figs. 16-20.
Armina abbotti Thompson, Cattaneo & Wong 1990: 403.

Description: Background color variable from gray to black, with a series of longitudinal ridges on the dorsum colored with white or yellow pigment. Ridges separated from each other, particularly in the center of the dorsum. Rhinophores large, posteriorly bordered by a white, transverse thickening of the mantle. Foot as wide as the mantle.

Geographic Range: North Carolina, South Carolina, Georgia, Florida, Texas, Mexico.

Family PROCTONOTIDAE Gray, 1853

Diagnosis: Body elongate, depressed, with numerous, slender, sometimes tuberculate cerata. Cerata with or without branches of the digestive gland, situated on the sides of the body and around the anterior margin of the head. Anterio-lateral corners of the velum with cylindrical tentacles. Rhinophores non-retractile, lamellate, irregularly rugose, or smooth. *Janolus* with a caruncle (or elevated ridge) between the bases of the rhinophores. Jaws large, wing-like, with a few denticles on the masticatory border in some cases. Radula with a normally smooth rachidian tooth and several smooth or denticulate laterals on each side. Anus dorso-posterior, in the midline of the body.

Biology: Proctonotids feed on bryozoans. Species of *Janolus*, the only genus represented in the Atlantic, eat arborescent species. When disturbed many species autotomize the cerata, which are sticky and adhere to anything that they come in contact with.

Janolus comis
Er. Marcus, 1955

Size: Up to 9 mm long

photos © L. Ianniello & © C. Redfern - left: Florida (LI); right: Abaco, Bahamas (CR)

Janolus mucloc
(Er. Marcus, 1958)

Size: Up to 26 mm long

photo © Er. Marcus - Brazil

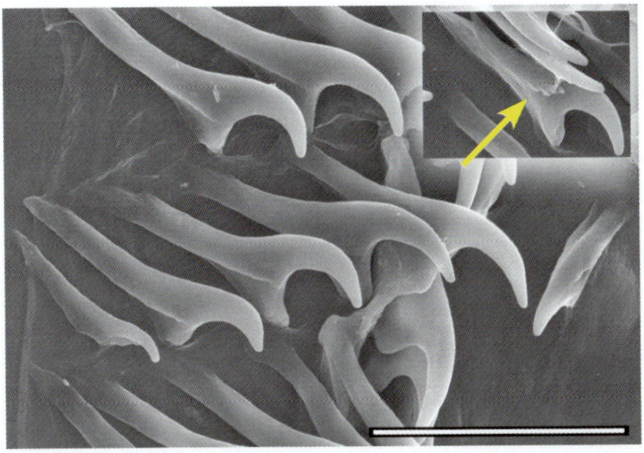

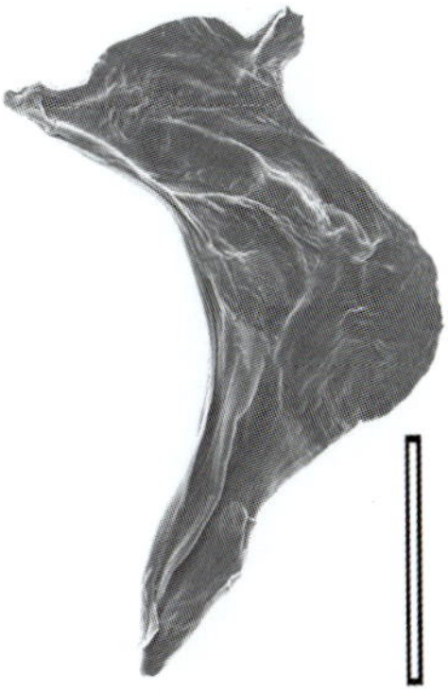

photos A. Valdés - SEM photos of *Janolus comis*; left: radular teeth with a detail of the inner side denticulation (arrow), scale bar = 30 µm; right: jaw, scale bar = 200 µm

Janolus comis Er. Marcus 1955: 170-173, fig. 226-236.
Janolus costacubensis Ortea & Espinosa 2000a: 80-82, figs. 1A-E, 2A-C.

Description: Background color translucent gray, peppered with opaque white spots. Foot wide, with the posterior end very elongate. Cerata all around the body, covered by small tubercles, normally elongate, occasionally swollen and balloon-like. Rhinophores long, irregularly lamellate, with a caruncle in between.

Geographic Range: Florida, Costa Rica, Cuba, Curaçao, Bahamas, Brazil.

Remarks: *Janolus costacubensis* was described as lacking denticles on the radular teeth, but examination of specimens assigned to this species confirms the presence of denticles on the inner side of the teeth.

Antiopella mucloc Er. Marcus 1958: 37-41, figs. 62-71.

Description: Background color translucent, almost transparent, with a longitudinal opaque white line running from in front of the rhinophores to the posterior end of the foot. Rhinophores lamellate, translucent, with white tips and a conspicuous caruncle in between. Cerata all around the body, elongate, with several longitudinal opaque white lines, a bright orange subapical spot and a translucent apex.

Geographic Range: Florida, Brazil.

Family DOTIDAE Gray, 1853

Diagnosis: Body small, elongate. Head with an oral velum that extends laterally. Rhinophoral sheaths high, smooth, jagged or with frontal expansions. Rhinophores smooth. Cerata tuberculate, containing branches of the digestive gland, arranged dorso-laterally on each side of the body. Ceratal tubercles usually arranged in rows or bands around the cerata. Some species with small, simple gills on the inner sides of the larger cerata. Anus and renal opening dorso-lateral in the inter-hepatic space. Jaw plates enclosing the buccal bulb. Radula with a single row of teeth, which have a single, large median cusp usually with a few flanking denticles.

Biology: Members of the Dotidae are suctorial feeders on hydroids. They bore holes through the hydroid perisarc to suck out the animal's fluids.

Remarks: The Caribbean species of *Doto* are in need of review and many of the identifications here proposed require anatomical verification. There are serious disagreements between authors on the identity of some species.

Doto pygmaea
Bergh, 1871

Size: Up to 10 mm long

photos J. Hamann - off Bell Channel, Bahamas; left: adult specimen; right: rhinophores

Doto pita
Er. Marcus, 1955

Size: Up to 7 mm long

photos J. Hamann - left: Isla Contoy, Mexico; right: egg masses, Vieques, Puerto Rico

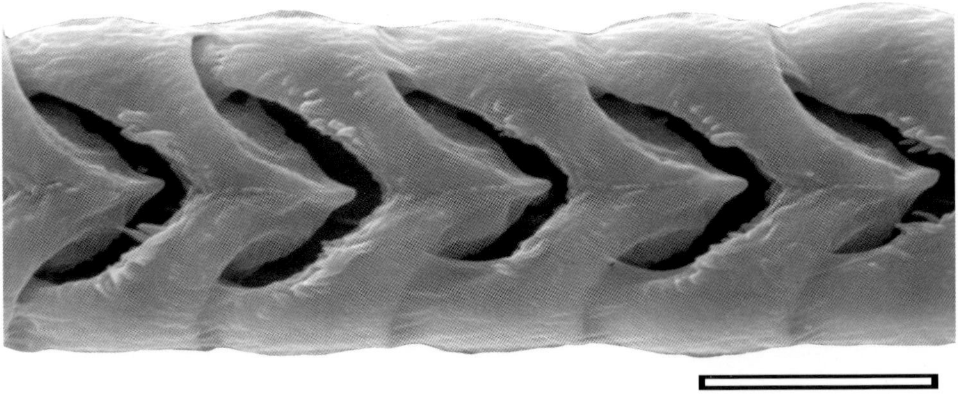

photo A. Valdés - SEM photos of the radular teeth of *Doto* sp. 3, scale bar = 10 μm

Doto pygmaea Bergh 1871: 1277-1280, pl. 11, figs. 1-10.
Doto doerga Ev. Marcus & Er. Marcus 1963: 39-41, figs. 47-51.

Description: Background color translucent gray, almost transparent, with numerous black spots all over the dorsum. Digestive gland white to red penetrating the cerata. Cerata with very elongate, club-shaped tubercles, translucent. Rhinophores with a few dark dots.

Geographic Range: Sargasso Sea, Cuba, Bahamas.

Natural History: Found on floating *Sargassum*.

Doto pita Er. Marcus 1955: 169-170, figs. 223-225.
?*Doto iugula* Ortea 2001a: 26-27, fig. 11, pl. 2, fig. J.

Description: Background color translucent gray, almost transparent, with numerous opaque white spots and some light brown patches, lined up on the center of the dorsum. Digestive gland white penetrating the cerata. Cerata with irregular tubercles having an opaque white gland inside and some superficial brownish spots. Rhinophores with opaque white dots.

Geographic Range: Amphiatlantic; Western Atlantic: Florida, Mexico, Costa Rica, Bonaire, Cuba, Puerto Rico, St. Kitts, Antigua, Barbados, Brazil.

Natural History: The egg mass is a long twisted ribbon of white eggs.

Remarks: *Doto iugula* is morphologically identical to *D. pita* except for the presence of orange spots on rhinophores in some specimens, which is probably due to intraspecific variation.

Doto uva
Er. Marcus, 1955

Size: Up to 11 mm long

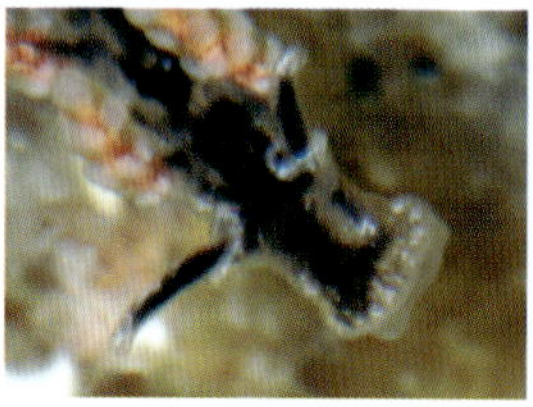

photos J. Hamann & A. Valdés - left: Guanaja, Puerto Rico (JH); right: Costa Rica (AV)

Doto divae
Ev. Marcus & Er. Marcus, 1960

Size: Up to 10 mm long

photos J. Hamann & T. Gosliner - left: La Parguera, Puerto Rico (JH); right: Sarasota, Florida (TG)

Doto wildei
Er. Marcus & Ev. Marcus, 1970

Size: Up to 8 mm long

photo J. Hamann - Spanish Harbor, Curaçao

Doto uva Er. Marcus 1955, p. 167-169, fig. 218-222.
?*Doto curere* Ortea 2001a: 17-18, fig. 7, pl. 2, fig. C.

Description: Background color dark, almost black. Cerata with a pinkish digestive gland and large, bluish, rounded tubercles. Rhinophores with a dark central axis and some opaque white dots. Rhinophoral sheath translucent, with a bluish edge and black pigment in the middle; a long, triangular extension emerges from the frontal end and curves upwards.

Geographic Range: Florida, Mexico, Cuba, Puerto Rico, Jamaica, Brazil.

Remarks: Ortea (2001) considered this species not present in the Caribbean. The specimen illustrated here matches the original description from Brazil as well as Ortea's (2001) record of *Doto chica*. *Doto curere*, from Costa Rica, is also very similar in shape and coloration and could constitute a synonym, although the ceratal tubercles appear to be more elongate.

Doto divae Ev. Marcus & Er. Marcus 1960a: 165-167, figs. 54-57.

Description: Background color translucent gray, with dense black pigment and numerous opaque white spots all over the dorsum. Cerata with a yellowish digestive gland and large, bluish, elongate tubercles. Rhinophores with a dark central axis and occasionally with a single lamella. Rhinophoral sheath translucent, with some opaque white spots and a jagged, opaque white edge.

Geographic Range: Florida, Puerto Rico, Bonaire, St. Lucia, Brazil.

Doto wildei Er. Marcus & Ev. Marcus 1970: 77-78, figs. 138-141.
?*Doto awapa* Ortea 2001a: 21-23, fig. 9A-E, pl. 2, fig. G, pl. 3, fig. D.

Description: Background color translucent gray, with opaque white spots and some brown pigment all over the dorsum. Cerata large, with light orange branches of the digestive gland and large, white, rounded tubercles. Rhinophores with opaque white spots. Rhinophoral sheath translucent, with some opaque white spots.

Geographic Range: Curaçao.

Remarks: Originally described as a subspecies of *Doto caramella*, we believe it is a distinct species. *Doto awapa*, from Costa Rica, is very similar in shape and coloration and could be a synonym.

photo © C. Redfern - Abaco, Bahamas

Doto chica
Ev. Marcus & Er. Marcus, 1960

Size: Up to 10 mm long

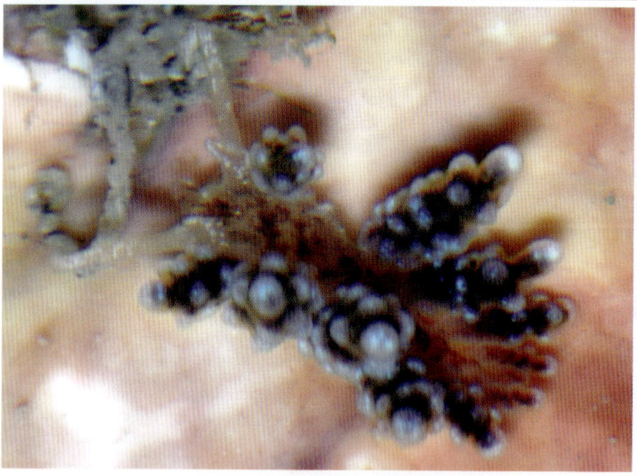

photo A. Valdés - Puerto Vargas, Costa Rica

Doto escatllari
Ortea, Moro, & Espinosa, 1997

Size: Up to 5 mm long

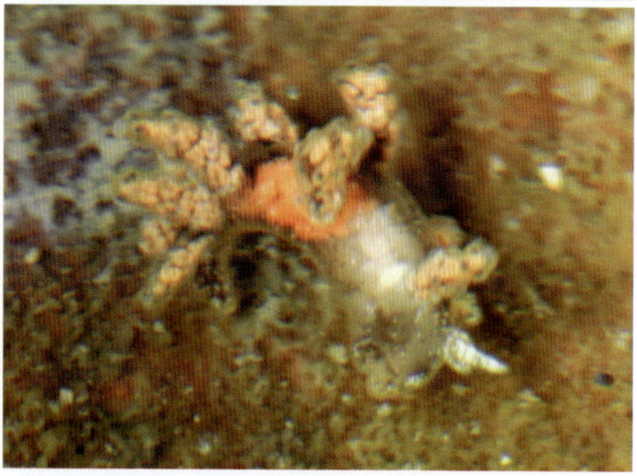

photo A. Valdés - Manzanillo, Costa Rica

Doto duao
Ortea, 1955

Size: Up to 6 mm long

Doto chica Ev. Marcus & Er. Marcus 1960a: 167-169, figs. 38-39.

Description: Background color translucent gray with brown pigment and opaque white spots on the dorsum. Cerata large, with elongate tubercles and a very large apical tubercle. Base of the cerata brown, digestive gland brown with white spots and ceratal tubercles with blue spots. Rhinophores and rhinophoral sheaths translucent with opaque white dots.

Geographic Range: Florida, Curaçao, Puerto Rico.

Remarks: *Doto umia*, originally described from Greenland, is not a synonym. Ortea's (2001) record of this species has triangular extensions on the rhinophoral sheaths and lack elongate tubercles on the tips of the cerata (see remarks of *Doto uva*). The specimen illustrated here matches the original description of the species.

Doto escatllari Ortea, Moro, & Espinosa 1997:131-132, fig. 3A-D, pl. 1, figs. D-E.

Description: Background color translucent gray with a series of dark brown spots on the dorsum. Cerata large, with dark brown branches of the digestive gland and bluish, rounded tubercles. Apical tubercles much larger than the rest. Rhinophores and rhinophoral sheaths translucent with opaque white dots. Sheaths with small frontal extensions.

Geographic Range: Amphiatlantic; Western Atlantic: Costa Rica, Barbados.

Doto duao Ortea 2001a: 28-30, fig. 12A-D, 13, pl. 1, fig. A, pl. 2, fig. K, pl. 3, fig. H.

Description: Background color translucent gray, almost transparent, with a series of opaque white spots on the dorsum. Digestive gland visible as an orange mass. Cerata with orange branches of the digestive gland and translucent tubercles with blue glands inside. Rhinophores and rhinophoral sheaths translucent with opaque white dots more densely arranged on the sheath edge and rhinophore.

Geographic Range: Costa Rica.

Doto sabuli
Ortea, 2001

Size: Up to 5 mm long

photo © T. Gosliner - Lake Surprise, Florida

Doto varaderoensis
Ortea, 2001

Size: Up to 10 mm long

photo J. Hamann - Culebra Island, Puerto Rico

Doto cabecar
Ortea, 2001

Size: Up to 14 mm long

photos J. Hamann - left: Cooper Island, Virgin Islands; right: rhinophores, Culebra, Puerto Rico

Doto sabuli Ortea 2001a: 9-10, fig. 2A-F, pl. 2, fig. A, pl. 3, fig. F.

Description: Background color translucent gray, with dark brown or purple pigment and numerous opaque white spots on the dorsum. Digestive gland visible. Cerata large, translucent, with dark red spots at the base. Branches of the digestive gland visible, cream. Ceratal tubercles short with pale blue spots inside. Gills conspicuous. Rhinophores with one lamella, covered with opaque white and dark dots. Rhinophoral sheath with blue spots at the edge.

Geographic Range: Mexico, Bahamas.

Doto varaderoensis Ortea 2001a: 37-39, fig. 18A-E, pl. 1, fig. D.

Description: Background color translucent gray, with dark pigment and numerous opaque white spots on the dorsum. Digestive gland visible as an orange mass. Cerata large, with yellowish white branches of the digestive gland and translucent, rounded tubercles with pale blue spots inside. Gills conspicuous. Rhinophores covered with opaque white dots. Rhinophoral sheath with a blue edge and an anterior triangular extension.

Geographic Range: Cuba, Puerto Rico.

Doto cabecar Ortea 2001a: 34-37, fig. 16A-C, 17A-D, pl. 2, fig. M, pl. 3, fig. L-M.

Description: Background color bright red, dorsum covered with opaque white dots. Cerata with yellow branches of the digestive gland and large, rounded, translucent tubercles with a bright blue gland inside. Rhinophores and rhinophoral sheaths translucent with numerous opaque white dots. Rhinophoral sheaths with frontal triangular extensions.

Geographic Range: Puerto Rico, Virgin Islands.

Remarks: The specimens illustrated here differ from the original description of *Doto cabecar* by lacking dark brown pigment on the dorsum.

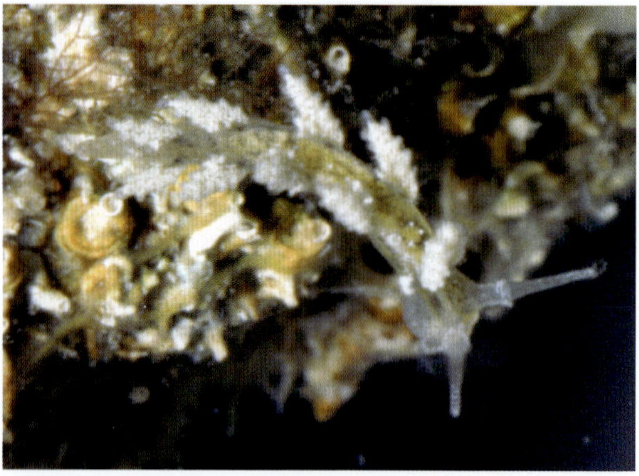

Doto sp. 1

Size: Up to 5 mm long

photos A. Valdés & J. Hamann - left: Key Largo, Florida (AV); right: Martinique (JH)

Doto sp. 2

Size: Up to 9 mm long

photo J. Hamann - Pidgeon Island, St. Lucia

Doto sp. 3

Size: Up to 6 mm long

photos A. DuPont - Bahamas; left: side view of specimen; right: detail of the head

Description: Background color translucent gray, with some brown pigment all over the dorsum. Cerata small, with opaque white branches of the digestive gland and bluish, small, elongate tubercles. Rhinophores and rhinophoral sheath translucent, almost transparent, with some opaque white spots.

Geographic Range: Florida, Martinique.

Remarks: This appears to be an undescribed species.

Description: Background color translucent dirty yellow with brown pigment and opaque white spots on the dorsum. Cerata elongate, with rounded tubercles all similar in size. Base of the cerata brown, digestive gland yellow and ceratal tubercles with a single purple spot. Rhinophores and rhinophoral sheaths translucent with dark pigment. Rhinophores with a single lamella.

Geographic Range: Panama.

Remarks: This appears to be an undescribed species.

Description: Background color translucent gray with dense dark brown-purple pigment on the dorsum and sides of the body. Cerata with dark brown branches of the digestive gland and opaque white, small, conical tubercles. Rhinophoral sheaths translucent with some brown spots. Rhinophores with a small lamella near the base, dark brown pigment and opaque white dots.

Geographic Range: Bahamas.

Remarks: This appears to be an undescribed species.

Family TRITONIIDAE Lamarck, 1809

Diagnosis: Body elongate, with the edge of the mantle bearing a series of branched secondary gills resembling cerata. Mantle margin normally reduced and body relatively high in profile. Oral veil usually bilobed, bearing papillae, often branched, along the edge. Rhinophoral sheath usually high with a lobed or scalloped edge. Rhinophores with a smooth stalk and a characteristic terminal club formed of a ring of vertical papillae. Anal, reproductive and renal openings on the right side of the body below the mantle edge. A pair of large, heavily chitinized jaw plates enclose the anterior end of the muscular buccal bulb. Radula composed of a median, often tricuspidate tooth and many pointed lateral teeth. First lateral tooth on each side normally differentiated from the rest. Some species with chitinous plates in the stomach, forming a ring around the entrance to the intestine.

Biology: Tritoniids feed on alcyonarian soft corals or gorgonians. Many species are very cryptic and mimic the shape and color of the soft corals on which they feed. The gills or cerata often closely resemble the soft coral polyps.

photo J. Hamann - Tortola, Virgin Islands

Tritonia bayeri

Ev. Marcus & Er. Marcus, 1967

"Crisscross tritonia"

Size: Up to 11 mm long

photo J. Hamann - Turneffe Reef, Belize

Tritonia hamnerorum

Gosliner & Ghiselin, 1987

"Bill and Peggy Hamner's tritonia"

Size: Up to 15 mm long

photos A. Valdés - SEM photos of *Marionia tedi*; left: rachidian teeth, scale bar = 100 µm; center left: outer lateral teeth, scale bar = 50 µm; center right: jaw, scale bar = 500 µm; right: detail of the jaw masticatory border, scale bar = 3 µm

Tritonia bayeri Ev. Marcus & Er. Marcus 1967a: 101-1-4, figs. 131-134.
Tritonia bayeri misa Ev. Marcus & Er. Marcus 1967c: 211-212, figs. 13-14.

Description: Background color translucent gray with a very conspicuous reticulate network of opaque white. Edge of the oral veil with four relatively long and large appendages. Cerata relatively small, branched.

Geographic Range: Georgia, Florida, Belize, Honduras, Cayman Islands, Virgin Islands, Guadeloupe, Barbados.

Natural History: Inhabits reefs down to 77 m depth, feeds on gorgonians

Tritonia hamnerorum Gosliner & Ghiselin 1987: 428-436, figs. 1-11.

Description: Background color white with a pale gray dorsum. A series of irregular, longitudinal thin white lines run along the length of the body. Rhinophoral sheaths white. Cerata short, branched.

Geographic Range: Florida, Mexico, Belize, Bahamas, Cayman Islands.

Natural History: Feeds on the gorgonian sea fans *Gorgonia ventilina* and *Gorgonia flabellum*. Dense populations with as many as 1,700 nudibranchs have been reported on a single colony of *G. ventilina*. This species takes distasteful chemicals from the sea fans and stores them for its own defense.

Tritonia wellsi
Ev. Marcus, 1961

"Wells' tritonia"

Size: Up to 10 mm long

photos J. Hamann - left: Ranguana & Lark Cayes, Belize; right: egg mass, Tortola, Virgin Islands

Tritoniopsis frydis
Er. Marcus & Ev. Marcus, 1970

Size: Up to 20 mm long

photos J. Hamann - Left: orange color form, Curaçao; right: white color form, Turneffe, Belize

Marionia tedi
Ev. Marcus, 1983

Size: Up to 40 mm long

photo © S. Hourdez - Mississippi Canyon, Louisiana

Tritonia wellsi Er. Marcus 1961: 146-148, figs. 22-24.

photo J. Hamann - ventral view, Roatán, Honduras

Description: Background color translucent pale gray, with numerous, small, opaque white dots on the dorsum. Branching cerata.

Geographic Range: North Carolina, Florida, Belize, Honduras, Costa Rica, Virgin Islands, St. Maarten/St. Martin, Brazil.

Natural History: Inhabits reefs and feeds on gorgonians.

Tritoniopsis frydis Er. Marcus & Ev. Marcus 1970: 75-77, figs. 134-137.

photo J. Hamann - three color forms, Ship Rocks, Bahamas

Description: Background color from translucent white to orange, with a pattern of opaque patches. The color difference does not appear to be food related as different color forms can be found together on the same colony of the sea fans on which they feed. Cerata highly branched, rhinophores relatively small.

Geographic Range: Florida, Belize, Honduras, Bermuda, Bahamas, Cayman Islands, Curaçao.

Natural History: Feeds on gorgonian sea fans of the genus *Plexaurella*.

Marionia tedi Ev. Marcus 1983: 200, 203-207, figs. 48-74.

Description: Background color uniformly translucent white with a pinkish tinge. Viscera visible through the skin as an opaque white mass. Rhinophores, cerata and velum having the same color as the rest of the body. Velum bilobed with about six processes on each lobe.

Geographic Range: Louisiana, Florida, Venezuela, Mexico.

Natural History: Found between 60-624 m depth in different environments, including hydrocarbon cold seeps. Probably feeds on the gorgonian sea fan *Callogorgia americana*.

Family BORNELLIDAE Bergh, 1874

Diagnosis: Body narrow and elongate, mantle margin completely absent. Several pairs of dorsal papillae are present along the dorso-lateral margins of the body. Cerata-like papillae sometimes branched and usually with 1-2 gills attached medially near the base. Oral tentacles palmate, with 5-10 pointed papillae in a single or double row. Rhinophoral sheath high, usually branched, very similar in shape to the paired dorsal papillae along the body. Rhinophores lamellate. Anal, reproductive, and renal openings ventral to the dorsal papillae on the right anterior quarter of the body. Buccal bulb with a thin cuticular lining on each side of the mouth composed of numerous rodlets. A pair of jaws enclose the buccal bulb. Radula with a prominent median tooth in each row, which has a large, heavily denticulate, median cusp. Lateral teeth narrow, elongate, sometimes with denticles. In the genus *Bornella*, a muscular chamber, lined with chitinous spines, separates the stomach from the intestine.

Biology: Members of the Bornellidae feed on hydrozoans. When disturbed some species are able to swim by lateral flexions of the body.

photo J. Hamann - Turneffe, Belize

Bornella calcarata
Mörch, 1863

“Tasseled nudibranch””

Size: Up to 65 mm long

photos J. Hamann - two adult specimens, Deshais, Guadeloupe

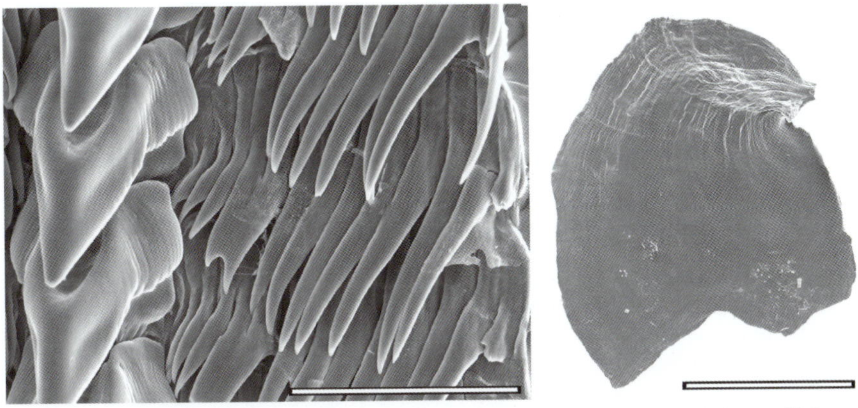

photos A. Valdés - SEM photos of *Bornella calcarata*; left: radular teeth, scale bar = 100 µm; right: jaw, scale bar = 40 µm)

Bornella calcarata Mörch 1863: 36-37.

Description: Background color translucent to opaque white with a reticulate pattern of orange or orange-red. In some specimens the orange color covers most of the animal leaving small, oval white areas. Tips of cerata highly branched.

Geographic Range: Mexico, Belize, Mexico, Costa Rica, Colombia, Cayman Islands, Jamaica, Puerto Rico, Virgin Islands, Guadeloupe, St. Lucia, St. Vincent & the Grenadines, Grenada, Trinidad & Tobago, French Guyana, Surinam, Brazil.

Natural History: Inhabits reefs. It feeds on hydroids. Animal swims by vigorous lateral flexions of the body when disturbed.

photos J. Hamann - Cozumel, Mexico - left: dorsal view of head; right: ventral view of head

Family LOMANOTIDAE Bergh, 1890

Diagnosis: Body elongate, with the edge of the mantle forming a sinuous ridge. Short pointed cerata lacking gills present along the edge of the ridge. Head with one or two pairs of oral tentacles. Rhinophores with fine transverse lamellae and a blunt rounded apex; they emerge from elevated sheaths with papillae around their edges. Anus opening just below the mantle ridge on the right side about one third of the way down the body, just posterior to the pericardium. Anterior corners of the foot rounded or tentacular. Jaws enclosing the buccal bulb with several rows of multi-denticulate plates along the masticatory border. Radula lacking a median tooth. Innermost tooth on each side short. A long pointed cusp emerges from a broad blade which is denticulate on both sides. Other lateral teeth similar in shape, with a reduced central cusp. The blade length increases laterally along the row, reaching its maximum about three teeth from the end.

Biology: Species of Lomanotidae feed on a variety of hydroids. Most species are cryptic on their food source.

photo J. Hamann - Bahamas

Lomanotus vermiformis
Eliot, 1908

Size: Up to 40 mm long

photo J. Hamann - Sharp Rocks, Bahamas

Lomanotus phiops
Er. Marcus, 1957

Size: Up to 8 mm long

photos A. Valdés - SEM photos of *Lomanotus vermiformis*; left: radular teeth, scale bar = 30 µm; right: jaw plates, scale bar = 10 µm

Lomanotus vermiformis Eliot 1908: 88-90.
Lomanotus stauberi Clark & Goetzfried 1976: 474-478, fig. 1A-H.

Description: Background color pale brown with dark brown spots and opaque yellow lines. Cerata translucent white. Body extremely elongate and narrow, with a number of short and pointed cerata. Rhinophoral sheaths with papillae.

Geographic Range: Circumtropical; Western Atlantic: Florida, Bahamas.

Natural History: Feeds on the hydroid *Lytocarpus philippinus* on which it is very cryptic. It produces a sticky mucous thread when crawling and swims violently with vigorous lateral flexions of the body when disturbed.

Remarks: *Lomanotus stauberi* described from the Caribbean is a synonym of the circumtropical species *L. vermiformis*.

Lomanotus phiops Er. Marcus 1957: 447-451, figs. 155-163.

Description: Background color variable from translucent gray to pale brown, with white spots. There are conspicuous patches of brown towards the tips of the dorsal appendages and on the lobes of the rhinophoral sheaths. Rhinophoral sheaths divided into five small lobes; rhinophores lamellate with small cylindrical tips. Two ridges extend backwards from the rhinophores, decorated with triangular cerata of variable size.

Geographic Range: Honduras, Bahamas, Brazil.

Natural History: Found on floating algae.

Family SCYLLAEIDAE Alder & Hancock, 1855

Diagnosis: Body elongate with the dorso-lateral mantle ridge expanded into one or two large flattened lobes on each side. Inner (dorsal) side of each ceras with dendritic, tubular, transparent gills. Oral velum normally bilobed. Rhinophores with a few lamellae, emerging from large, flattened sheaths similar in shape to the cerata. Posterior end of the body with a medio-dorsal keel. Anus and renal duct open mid-way along the body, ventral to the single right ceras in genera with only one pair of or cerata, or between the two right cerata. Jaws partially enveloping the buccal bulb. Radula with several teeth, sometimes with a central tooth, which is an elongate plate bearing a single pointed cusp with four to five denticles on each side. Lateral teeth are more elongate, slightly asymmetrical. Junction of the intestine and the stomach with a band of chitinous plates.

Biology: Scyllaeids feed on epiphytic hydroids growing on brown algae or sea grasses. All species can swim using lateral flexions of the body assisted by the erection and flattening of the rhinophoral sheaths and cerata to increase the area of the body. Some genera are strictly benthic whereas others are associated with drifting, oceanic brown algae.

photo J. Hamann - Bahamas

Scyllaea pelagica
Linnaeus, 1758

"Sargassum nudibranch"

Size: Up to 102 mm long

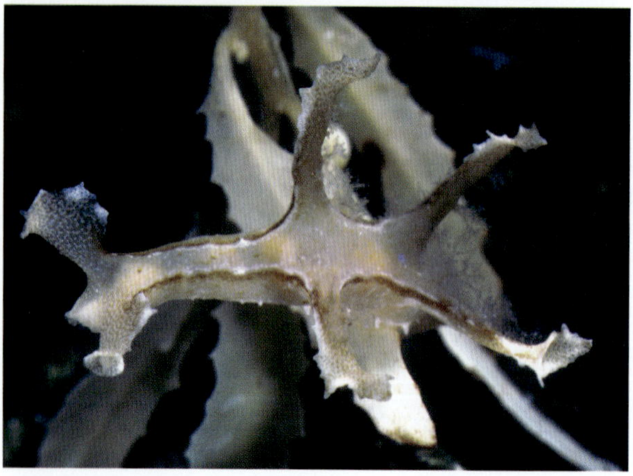

photo J. Hamann - Bahamas

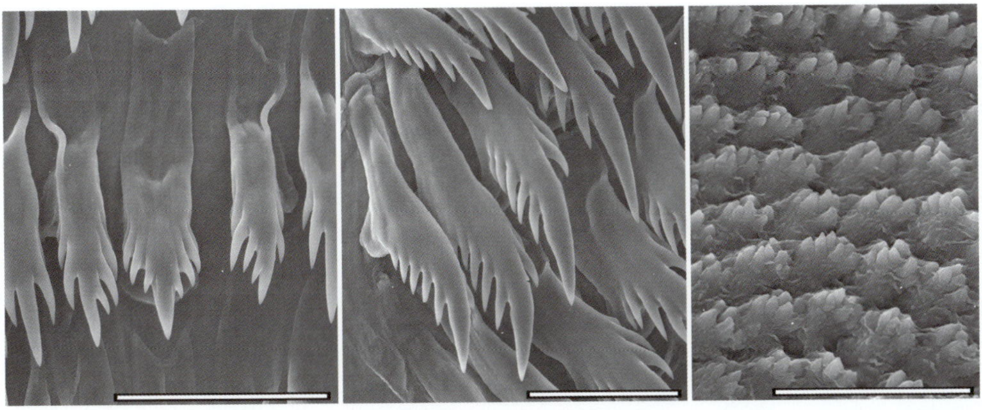

photos A. Valdés - SEM photos of *Crosslandia* cf. *daedali*; left: inner teeth, scale bar = 50 µm; center: outer teeth, scale bar = 50 µm; right: jaw plates, scale bar = 20 µm

---

Scyllaea pelagica Linnaeus 1758: 656.

photo A. DuPont - Bahamas

Description: Background color green to yellow with brown mottling, and some white markings, matching the *Sargassum* weed on which it lives. A row of brilliant blue spots may be present on each side of the body. Body flattened dorso-ventrally with large elaborate flattened rhinophoral sheaths. Two pairs of dorso-lateral lobes. Posterior end of the body with a flattened medio-dorsal crest that stands vertically. Small scattered conical tubercles on the sides of the body and fine dendritic gills over the inner surface of the lobes and dorsum.

Geographic Range: Cosmopolitan in temperate and warm waters; Western Atlantic: Massachusetts, North Carolina, Georgia, Florida, Texas, Costa Rica, Bonaire, Bermuda, Bahamas.

Natural History: This species normally lives pelagicaly on pieces of floating *Sargassum* seaweed where it feeds on attached hydroids. Sometimes found on benthic *Sargassum*.

photo J. Hamann - Soutiere, St. Lucia

Notobryon cf. wardi
Odhner, 1936

Size: Up to 62 mm long

photo J. Hamann - Soutiere, St. Lucia

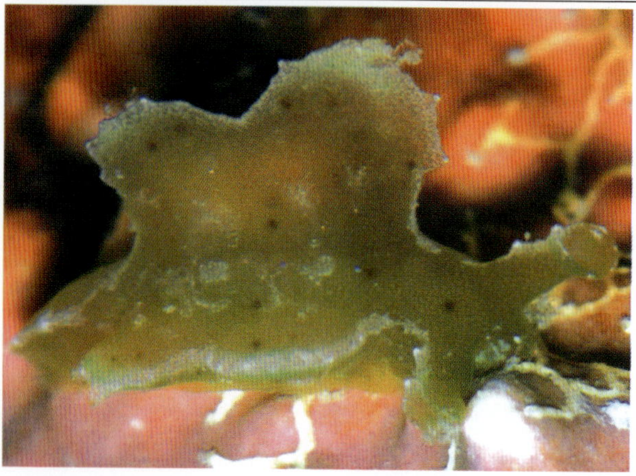

photo A. Valdés - Manzanillo, Costa Rica

Crosslandia cf. daedali
Poorman & Mulliner, 1981

Size: Up to 25 mm long

Notobryon wardi Odhner 1936: 1099-1103, figs. 31-38, pl. 1, figs. 1-3.

Description: Background color translucent gray with some pale brown and opaque white irregular pigment. Two pairs of dorso-lateral lobes with conspicuous dendritic gills; posterior end of the body with a flattened, medio-dorsal crest which stands vertically.

Geographic Range: Tropical Indo-Pacific and Western Atlantic: Honduras, St. Lucia, Virgin Islands, St. Vincent & the Grenadines.

Remarks: This is the first record of *Notobryon* from the Atlantic Ocean. The identification as the tropical Indo-Pacific *Notobryon wardi* is provisional. This species could be undescribed and it is possible that more than one species are illustrated here.

photos J. Hamann - left: Virgin Islands; right: St. Vincent

Crosslandia daedali Poorman & Mulliner 1981: 96-99, figs. 1-6.

Description: Background color greenish brown with some darker brown spots and a few bright blue dots. Two pairs of lateral lobes fused together at the base, the anterior longer than the posterior. Gills retractable.

Geographic Range: Eastern Pacific and Western Atlantic: Costa Rica.

Natural History: The identification of this species as the tropical eastern Pacific *Crosslandia daedali* is provisional. This is the first record of *Crosslandia* from the Atlantic Ocean.

Families HANCOCKIIDAE MacFarland, 1923 & TETHYDIDAE Rafinesque, 1815

Diagnosis of Hancockiidae: Body elongate with a long narrow foot. No distinct oral veil; each side of the mouth with a rounded palmate oral tentacle. Rhinophores with a few vertical lamellae, emerging from an elevated papillate sheath. Mantle margin reduced to an inconspicuous ridge, at regular intervals large compound cerata rise from this ridge. Each ceras consists of a cluster of long, tapering, papillae arranged in a circle with a common base. Anal, renal and genital openings ventral to the cerata on the right anterior part of the body. Buccal bulb with a thin cuticular lining on each side of the mouth composed of numerous rodlets. A pair of jaws enclose the buccal bulb. Rachidian tooth with a large median cusp having denticles on each side. Lateral teeth usually smooth, irregular triangular plates. Digestive gland divided into three lobes.

Biology of Hancockiidae: Species of Hancockiidae feed on hydroids and store nematocysts from their prey in their cerata. The nematocysts are discharged when the nudibranch is disturbed.

photo J. Hamann - Chimana Grande, Venezuela

Hancockia rycia
Er. Marcus, 1957

"White-spotted hancockia"

Size: Up to 15 mm long

photo J. Hamann - Guanaja, Honduras

***Melibe* sp.**

Size: Up to 13 mm long

Diagnosis of Tethydidae: Body elongate, with a large oral hood on the mouth area, edged with tentacles. Rhinophores having horizontal lamellae, emerging from cylindrical or flattened sheaths. Cerata usually lacking digestive gland branches, arranged in a single row along each side of the body, and filled with fluid. Branched gills on the inner side of the cerata, sometimes present on the outer side of the cerata and on the dorsum. Anal, genital and renal openings ventral to the cerata. Radula absent, replaced by chitinous thickenings on each side of the mouth in some species, and true jaws in others. Stomach usually with a band of chitinous plates.

Biology of Tethydidae: When disturbed, the larger species are able to swim by lateral flexions of the body aided by the erection of the cerata to increase the surface area. In extreme cases tethydids can autotomize the cerata, which can later regenerate. They feed by capturing small crustaceans in their large extensible oral hood.

Remarks: The family Tethydidae contains only two genera: *Tethys* and *Melibe*. The species *Tethys occidentalis* (Odhner, 1936) was described from Dominica Island in the Caribbean Sea, but never collected again. Below is the first record of the genus *Melibe* for the western Atlantic.

Hancockia rycia Er. Marcus 1957: 451-454, figs. 164-174.

Description: Background color translucent gray, with yellow internal organs showing through the skin and the entire dorsal surface including cerata. Rhinophoral sheaths and frontal lobes covered with small white dots. Cerata aligned along either side of the dorsum, with the bases densely pigmented with opaque white. Rhinophores with a bulbous tip, surrounded by short finger-like extensions from sheath rim. Two distinctive lobes at the front of the head, each with several finger-like processes.

Geographic Range: Venezuela, Bahamas, Curaçao, Barbados, Brazil.

Description: Background color translucent gray, almost completely transparent, with the internal organs visible as opaque white structures. Additional opaque white and brown spots present on body and cerata. Body elongate, covered by small conical tubercles, with four flat, tuberculate cerata on each side. Oral hood large, rounded, with a series of papillae on its edge. Rhinophoral sheaths elevated with short papillae on the edge.

Geographic Range: Honduras, Puerto Rico, Virgin Islands.

Remarks: This appears to be an undescribed species. It constitutes the first record of the genus *Melibe* in the Caribbean Sea and the western Atlantic.

Family FLABELLINIDAE Bergh, 1889

Diagnosis: Body elongate, narrow and tapering. Cerata usually arranged in single rows or clusters of rows. Cerata insert individually into the body or each row arises from a more or less raised lobe. Rhinophores smooth, papillate or lamellate. Oral tentacles usually long and tapering with a broad triangular base. Foot usually broad with tentaculiform anterior corners. Anus opens either below the ceratal rows or in the inter-hepatic space (acleioproctic). Jaw with several rows of irregular rodlets along the masticatory border. Radula triseriate (three teeth per row); central tooth with a raised cusp to a median cusp, flanked with denticles, lateral teeth normally denticulate along the inside edge.

Biology: Members of the Flabellinidae feed on hydroids. Most species display bright external coloration.

photo A. Valdés - Sanibel Island, Florida

Flabellina cf. *bandeli*
(Ev. Marcus, 1976)

Size: Up to 15 mm long

photos J. Hamann - Virgin Islands

Flabellina verta
(Ev. Marcus, 1970)

Size: Up to 11 mm long

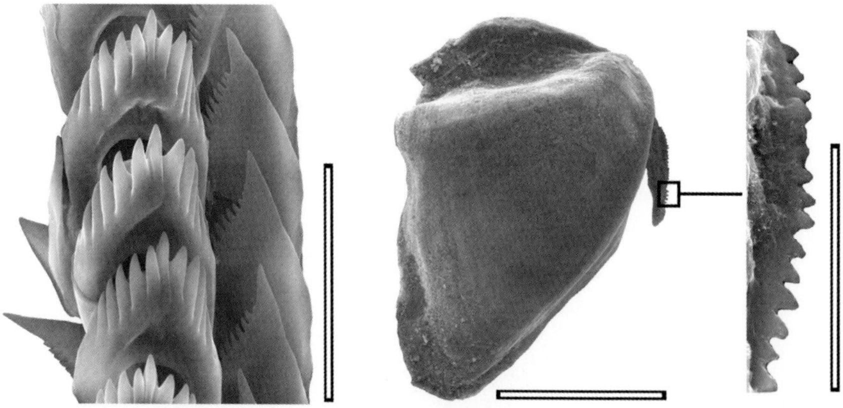

photos A. Valdés - SEM photos of *Flabellina* cf. *bandeli*; left: radular teeth, scale bar = 50 µm; center: jaw, scale bar = 300 µm; right: detail of the jaw masticatory border, scale bar = 50 µm

---

Calmella bandeli Ev. Marcus 1976a: 143-145, figs. 29-32.

Description: Background color bluish white. Three longitudinal opaque white lines run dorsally and laterally the entire length of the body; between the rhinophores the dorsal line splits into two lines that enter each oral tentacle. Ceratal wall translucent with some opaque white spots; digestive gland bright red. Rhinophores smooth. Cerata short. Posterior end of body short, triangular.

Geographic Range: Florida, Bahamas, Jamaica, Curaçao.

Natural History: Feeds on hydrozoans of the genus *Eudendrium*.

Remarks: The original description of *Calmella bandeli* is very incomplete, but mentions a white body with fluorescent shades and bright red cerata. The illustration represents a specimen of *Flabellina* with smooth rhinophores, short cerata and a triangular posterior end of the body.

Coryphella verta Ev. Marcus 1970: 211-213, figs. 9-15.

Description: Background color translucent white, with a broad opaque white or yellow dorsal central line that splits into two on the head area, and a thinner yellow line on each side of the body. Ceratal wall translucent; digestive gland pale brown, with opaque white or yellow dots on the cerata. Oral tentacles and rhinophores opaque white with translucent bluish bases. Rhinophores smooth.

Geographic Range: Virgin Islands, Brazil.

Natural History: Feeds on hydrozoans of the genus *Eudendrium*.

Remarks: The specimens illustrated here strongly resembles the original description of *Flabellina verta* and is therefore assigned to this species.

Flabellina dushia
(Ev. Marcus & Er. Marcus, 1963)

Size: Up to 20 mm long

photo J. Hamann - Bat Cave, Martinique

Flabellina marcusorum
Gosliner & Kuzirian, 1990

Size: Up to 25 mm long

photos J. Hamann - Curaçao; left: entire specimen; right: ventral view of head

Flabellina hamanni
Gosliner, 1994

Size: Up to 32 mm long

photos J. Hamann - Freeport, Bahamas; left: entire specimen; right: ventral view of head

Coryphella dushia Ev. Marcus & Er. Marcus 1963: 41-42, figs. 52-54.

Description: Background color opaque white, including oral tentacles and rhinophores. Ceratal wall translucent; digestive gland from red to bright orange-yellow; some animals with an orange or yellow digestive gland have a deeper orange or red band at the upper end of the digestive gland duct giving the cerata the appearance of having a subterminal orange-red band. Rhinophores smooth. Posterior end of body elongate.

Geographic Range: Florida, Bahamas, Jamaica, Curaçao, Martinique.

Natural History: Feeds on hydroids.

Flabellina marcusorum Gosliner & Kuzirian 1990: 4-9, figs. 1B, 4-5.

Description: Background color translucent pink or orange. Oral tentacles, foot corners, rhinophores and cerata similarly colored, with an opaque white or yellow distal part, followed by a broad purplish band. Cerata translucent with the branches of the digestive gland visible as a wide dark blue to brown strand. Rhinophores with up to 100 papillae, sometimes white, on the posterior side.

Geographic Range: Tropical Eastern Pacific and Atlantic; Western Atlantic: Costa Rica, Curaçao, Venezuela, Brazil.

Remarks: *Flabellina arveloi* Ortea & Espinosa 1998, from Cape Verde, is most likely a synonym.

Flabellina hamanni Gosliner 1994: 179-182, figs. 3B, 8-10.

Description: Background color translucent bluish. Oral tentacles, foot corners, rhinophores and cerata similarly colored, with an opaque white or yellow distal part, followed by a thin purplish band. Cerata opaque white, with the digestive gland not visible. Rhinophores with up to 100 papillae on the posterior side, sometimes brown or purple.

Geographic Range: Bahamas, Turks and Caicos.

Remarks: Very similar to *Flabellina marcusorum*, but distinguishable by having opaque white cerata with more rounded tips and the digestive gland branches not visible.

Flabellina engeli
Ev. Marcus & Er. Marcus, 1968

Size: Up to 25 mm long

photos J. Hamann - left: Punta Tigrillo, Venezuela, right: head, Pidgeon Island, St. Lucia

Flabellina sp. 1

Size: Up to 21 mm long

photo J. Hamann - Testigos, Venezuela

Flabellina sp. 2

Size: Up to 19 mm long

photo J. Hamann - Venezuela; left: full view of animal; right: rhinophores

Flabellina engeli Ev. Marcus & Er. Marcus 1968: 139-142, figs. 1-5.

Description: Background color translucent gray, with opaque white bands on each side, running between the ceratal clusters, and a white band along the dorsal mid-line of the posterior end. Orange or yellow triangular patches on each side of the body between ceratal clusters; a submarginal row of opaque white spots along the sides of the body. Cerata translucent with a brown or orange band about a third of the way down from the tip. Rhinophores lamellate, club-shaped, with white bands. Posterior end of body very elongate.

Geographic Range: Florida, Costa Rica, Colombia, Curaçao, Cuba, Puerto Rico, Barbados, St. Lucia, Martinique, Grenada.

Description: Background color translucent white, with a thin yellow dorsal central line that splits into two on the head area, and two thinner yellow lines on the sides of the body. Ceratal wall translucent showing the bright red digestive gland branches and the white cnidosac. Oral tentacles and rhinophores translucent white with opaque white pigment. Rhinophores smooth.

Geographic Range: Venezuela.

Remarks: This species is either undescribed or a color form of *Flabellina verta*.

Description: Background color translucent gray with yellow lines on each side of the dorsum, running between the ceratal clusters, and lateral yellow lines, one on each side, near the foot. Cerata translucent pale brown with white tips. Rhinophores club-shaped, lamellate. Posterior end of body short, triangular.

Geographic Range: Venezuela.

Remarks: This species is probably undescribed or a color form of *Flabellina engeli*. It is very similarly colored to typical *Flabellina engeli* but lacks white pigment on the dorsum, brown or orange bands on the cerata, white bands on the rhinophores, and the posterior end of the body is much shorter.

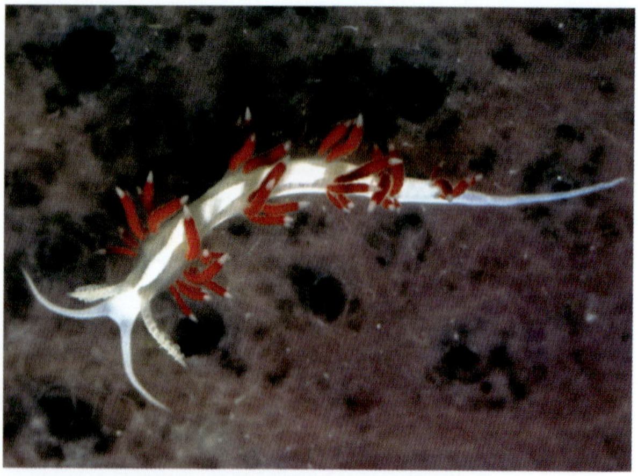

photo J. Hamann - Pidgeon Island, St. Lucia

Flabellina sp. 3

Size: Up to 12 mm long

photo © L. Ianniello - Riviera Beach, Florida

Flabellina sp. 4

Size: Up to 20 mm long

photos J. Hamann - Chimana Grande, Venezuela

Flabellina sp. 5

Size: Up to 16 mm long

Description: Background color translucent white with the head, oral tentacles and rhinophores opaque white. A wide central opaque white line, and two lateral lines, one on each side, run from the head to the posterior end of the body. Cerata with red branches of the digestive gland and a white cnidosac. Rhinophores annulate. Posterior end of the foot very elongate.

Geographic Range: Costa Rica, Curaçao, St. Lucia.

Remarks: This appears to be an undescribed species.

Description: Background color reddish orange with a central longitudinal purple line running from the anterior to the posterior end of the body. Rhinophores and oral tentacles with orange bases, deep purple distally, and having white apices. Cerata dark orange with opaque white cnidosacs.

Geographic Range: Florida.

Remarks: This appears to be an undescribed species.

Description: Background color translucent white, with a broad opaque white or yellow mid-dorsal line that splits into two on the head area, and a thinner yellow line on each side of the body. Ceratal wall translucent; digestive gland orange to red. Oral tentacles and rhinophores opaque white with translucent bluish bases. Rhinophores smooth.

Geographic Range: Venezuela.

Remarks: This species is either undescribed or a color form of *F. verta*.

Family EUBRANCHIDAE Odhner, 1934

Diagnosis: Body elongate, small, with relatively long, smooth rhinophores and short oral tentacles. Foot usually wider than the body and its anterior corners may be rounded, angular or tentaculiform. Few cerata, usually arranged in simple rows, inflated, sometimes with one or more rings of tubercles. Anus opens on the right side of the body in the high inter-hepatic space (acleioproctic), immediately behind the renal pore. Jaws with a masticatory border smooth or with a single row of teeth. Radula triseriate (with three teeth per row). Central tooth with four or five large denticles on each side of a pointed central cusp. Lateral teeth non-denticulate, with a wide base.

Biology: Members of the Eubranchidae feed on small colonial hydroids.

photo J. Hamann - Store Bay, Tobago

Eubranchus coniclus

(Er. Marcus, 1958)

Size: Up to 4 mm long

photo J. Hamann - Store Bay, Tobago

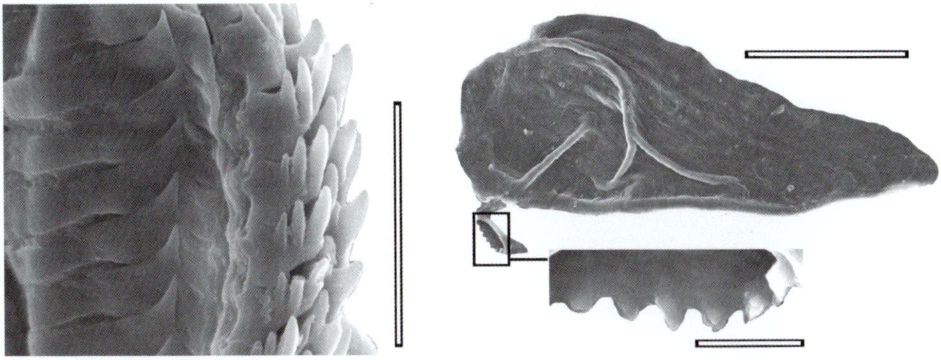

photos A. Valdés - SEM photos of *Eubranchus coniclus*; left: radular teeth, scale bar = 20 µm; upper right: jaw, scale bar = 100 µm; lower right: detail of the jaw masticatory border, scale bar = 10 µm

Capellinia conicla Er. Marcus 1958: 41-45, figs. 72-80.
Eubranchus convenientis Ortea & Caballer 2002: 81-85, figs. 2A-C, 3A-D, 6B, 7C.

Description: Background color translucent greyish with numerous white dots covering much of the body. Patches of brown or dark green sometimes occur on the posterior end of the foot and sides of the body, as well as one or two rings on the rhinophores. Cerata tuberculate, white or cream, spotted with greenish brown, and they alternate in size with the longest being about half the length of the body.

Geographic Range: Florida, Honduras, Costa Rica, Jamaica, Barbados, Tobago, Brazil.

Remarks: *Eubranchus convenientis* fits within the morphological variability of *E. coniclus*, thus these two species are regarded here as synonyms.

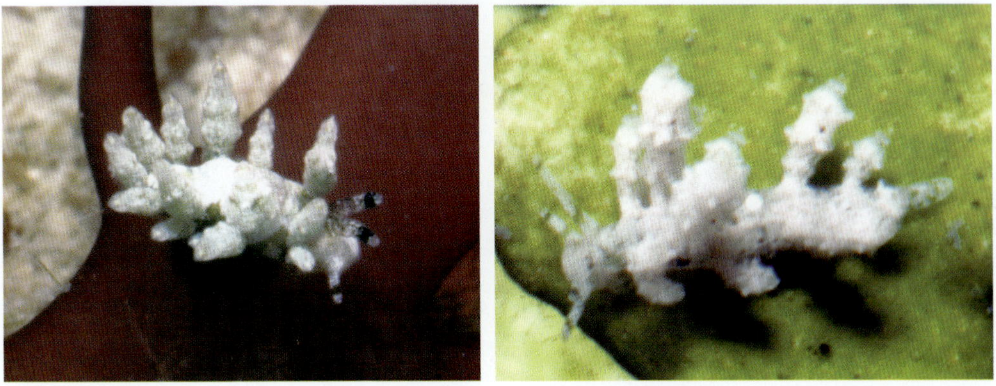

photos J. Hamann - left: Store Bay, Tobago; right: Guanaja, Honduras

photo J. Hamann - Pampatar, Venezuela

Eubranchus leopoldoi

Caballer, Ortea & Espinosa, 2001

Size: Up to 6 mm long

photo J. Hamann - Queen's Cove, Bahamas

***Eubranchus* sp.**

Size: Up to 8 mm long

photo J. Hamann - Garbage Hole, Bahamas

Unidentified aeolid

Size: Up to 10 mm long

Eubranchus leopoldoi Caballer, Ortea & Espinosa 2001: 55-56, fig. 14A-D, pl. 2, fig. E.

Description: Background color translucent gray with numerous red spots all over the body. Cerata inflated with a pattern of red and opaque, and a bluish band near the apex. Rhinophores and oral tentacles translucent with numerous opaque white spots and an orange band.

Geographic Range: Costa Rica, Bahamas.

Description: Background color greyish brown with some opaque white pigment on the head and the posterior end of the foot. Digestive gland visible through the body as a reddish mass connecting the bases of the cerata. Cerata translucent with visible digestive gland branches reddish and small brown tubercles; cnidosacs white.

Geographic Range: Bahamas.

Remarks: This appears to be an undescribed species.

Description: Background color translucent white with a number of subepidermical brown dots and epidermical opaque white spots all over the body. Cerata flat, having the same color as the rest of the body. Rhinophores and oral tentacles with opaque white apices.

Geographic Range: Bahamas.

Remarks: This appears to be an undescribed species. We are unable to determine the family placement for this species until anatomical information becomes available.

Family FACELINIDAE Bergh, 1889

photo A. Valdés - cerata

Diagnosis: Body elongate with a series of cerata arranged in rows, sometimes clusters of rows, or in arches. Rhinophores may be smooth, annulate, papillate or lamellate. Anus usually situated within the cerata group behind the inter-hepatic space (cleioproctic). Digestive gland with an uniseriate radula (a single row of teeth); radular teeth with long basal limbs and a large pointed cusp flanked by several prominent denticles.

Biology: Most facelinids feed on hydroids, except for species of the Indo-Pacific genus *Phyllodesmium* that feed on alcyonacean soft corals or gorgonians and species of *Favorinus* that feed on the eggs of other opisthobranchs. *Phyllodesmium* and *Favorinus* are the only two genera in the family that lack cnidosacs for nematocyst storage in the tips of their cerata. The nematocysts from alcyonaceans are possibly too

Austraeolis catina

Ev. Marcus & Er. Marcus, 1967

"Long horn aeolid"

Size: Up to 22 mm long

photos J. Hamann - left: Piton Bay, St. Lucia; right: ventral side of the head, Bahamas

Babakina anadoni

(Ortea, 1979)

"Gaudy aeolid"

Size: Up to 10 mm long

photo J. Hamann - Garbage Hole, Bahamas

small to be adequate for defense. When disturbed, several species of facelinids wave their cerata, aggressively pointing the tips containing the cnidosacs at the source of the attack.

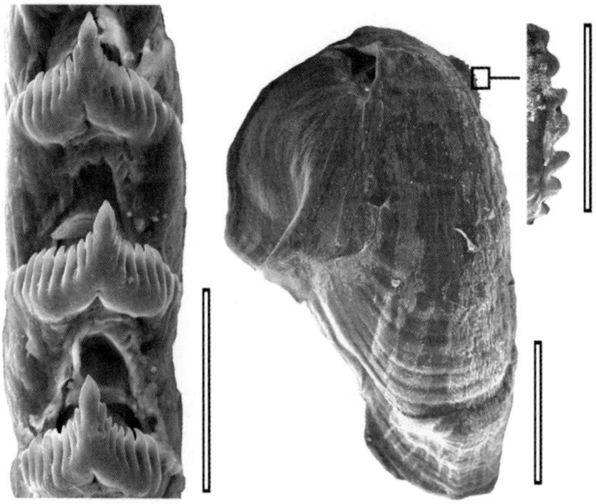

right photos A. Valdés - SEM photos of *Dondice occidentalis*; left: radular teeth, scale bar = 100 µm; center: jaw, scale bar = 500 µm; up and right: detail of the masticatory border of the jaw, scale bar = 100 µm

---

Austraeolis catina Ev. Marcus & Er. Marcus 1967a: 112-115.

Description: Background color translucent pale brown or greyish, with brown marks on the dorsum and white spots on the head. Cerata brown with some yellowish pigment, larger cerata with opaque white spots. Rhinophores lamellate.

Geographic Range: Florida, Belize, Puerto Rico, Bahamas, St. Lucia.

Natural History: Found down to 30 m depth.

Rioselleolis anadoni Ortea 1979: 416-418, figs. 1-9.

Description: Background color translucent gray with a bluish tinge. A conspicuous patch of opaque white at the center of the head, between the oral tentacles and the rhinophores. Oral tentacles purple on their lower half and white above. Cerata translucent grey with opaque white, yellow, blue and reddish pigment. Rhinophores lamellate, joined together for the first third of their length.

Geographic Range: Amphiatlantic; Western Atlantic: Bahamas, Grenada, Brazil.

Remarks: Cervera *et al.* (personal communication) confirmed that Caribbean populations of *Babakina* belong to the eastern Atlantic species *B. anadoni*.

Favorinus auritulus
Er. Marcus, 1955

"Long-eared favorinus"

Size: Up to 13 mm long

photos J. Hamann - left: Cave Point, Martinique; right: juvenile specimen, Store Bay, Tobago

---

Godiva rubrolineata
Edmunds, 1964

Size: Up to 21 mm long

photo J. Hamann - Guanaja, Puerto Rico

Noumeaella kristenseni
(Ev. Marcus & Er. Marcus, 1963)

"Frosty mordilla"

Size: Up to 16 mm long

photo J. Hamann - Guanaja, Honduras

Favorinus auritulus Er. Marcus 1955: 181-184, figs. 259-269.

Description: Background color translucent with scattered white patches all over the body and cerata. Bottom three quarters of rhinophores sometimes brown. Cerata orange to light brown. Rhinophores with one to three swellings.

Geographic Range: Florida, Curaçao, Bermuda, Bahamas, Jamaica, Puerto Rico, Antigua, Martinique, Tobago, Barbados, Bermuda, Brazil.

Natural History: Feeds on opisthobranch eggs, and the color of the eggs can change the color of the ceratal digestive gland.

Godiva rubrolineata Edmunds 1964: 23-27, figs. 14B, 15A-D, 16A-C.

Description: Background color translucent white with three longitudinal orange-red lines, one dorsal, two lateral. Cardiac area reddish. Rhinophores and oral tentacles with a translucent gray lower half and an opaque bluish white upper half. Cerata translucent with brown branches of the digestive gland and opaque white cnidosacs. Rhinophores smooth.

Geographic Range: Florida, Curaçao, Jamaica, Puerto Rico, Brazil.

Mordilla kristenseni Ev. Marcus & Er. Marcus 1963: 44-45, figs. 57-60.

Description: Background color translucent white or pale yellow brown with a network of opaque white lines on the dorsum. Cerata similarly colored, with an orange-brown spot just below each translucent tip and numerous white dots. Rhinophores tuberculate.

Geographic Range: Florida, Honduras.

Dondice occidentalis
(Engel, 1925)

"Western dondice"

Size: Up to 50 mm long

photos J. Hamann - Bat Cave Point, Martinique; left: dorsal view of specimen; right: ventral side of the head

photos J. Hamann & A. Valdés - left: Turtle Cove, Turks & Caicos (JH); right: Sarasota, Florida (AV)

Dondice parguerensis
Brandon & Cutress, 1985

Size: Up to 48 mm long

photo J. Hamann - Fajardo, Puerto Rico

Caloria occidentalis Engel 1925: 41-44, figs. 7-15.
Facelina coenda Er. Marcus 1958: 52-56, figs. 88-96.

Description: Background color translucent grayish. Head with a median line of variable width, from the anterior edge to behind the rhinophores; it ranges in color from red to orange or yellow. A white or blue broken line down the dorsal mid-line from behind the rhinophores to the posterior end of the body can sometimes be absent. A line on each side of the body from the head to the posterior end but usually broken beneath the ceratal groups; it ranges in color from red to yellow or blue. Cerata often with broad white or blue bands, covering the upper two thirds and sometimes a bright red or orange band near the apex. Rhinophores annulate, long.

Geographic Range: Florida, Mexico, Belize, Costa Rica, Colombia, Curaçao, Bonaire, Venezuela, Bermuda, Bahamas, Cayman Islands, Jamaica, Turks & Caicos, Grenada, St. Maarten/St. Martin, Martinique, Trinidad, Brazil.

Natural History: When disturbed this species sheds wiggly cerata.

photos J. Hamann - left: Florida; right: Hawksbill Creek, Bahamas

Dondice parguerensis Brandon & Cutress 1985: 139-144, figs. 1-2.

Description: Background color translucent grey with a longitudinal opaque white line in the middle of the dorsum. Cerata with a subapical white ring. Rhinophores and oral tentacles pigmented with opaque white on the distal portions. Rhinophores annulate.

Geographic Range: Puerto Rico, Panama.

Natural History: Found exclusively on the tentacles of the upside-down jellyfish *Cassiopea* in shallow mangrove areas.

Remarks: Differs from *Dondice occidentalis* in habitat and external morphology. The authors of this book do not agree on whether these two species are synonyms.

Nanuca sebastiani
Er. Marcus, 1957

"Orange spotted nanuca"

Size: Up to 12 mm long

photos J. Hamann & A. Valdés - left: Martinique (JH); right: St. Thomas, Virgin Islands (AV)

photo J. Hamann - Miami, Florida

Palisa papillata
Edmunds, 1964

Size: Up to 15 mm long

photo J. Hamann - Taino Beach, Bahamas

Pauleo jubatus
Millen & Hamann, 1992

"White-speckled aeolid"

Size: Up to 46 mm long

Nanuca sebastiani Er. Marcus 1957: 474-475, figs. 225-233.

photo J. Hamann - egg mass, Martinique

Description: Background color translucent greenish with numerous opaque white spots, and a series of areas with orange spots forming a blue or white cross-like pattern. Cerata with longitudinal opaque white lines and a white, narrow apex. Rhinophores annulate.

Geographic Range: Mexico, Costa Rica, Curaçao, Bonaire, Martinique, Cuba, Cayman Islands, Virgin Islands, Barbados, Bermuda, Brazil.

Natural History: Inhabits shallow areas with abundant algae and hydroids, and attached to benthic *Sargassum*.

Palisa papillata Edmunds 1964: 12-16, figs. 8A-E, 9A-C, 10A.

Description: Background color silvery gray, with the dorsal surface including cerata covered with numerous opaque white spots. Cerata translucent with the digestive gland duct pale blue and dark brown spots at the base. Rhinophores tuberculate.

Geographic Range: Florida, Jamaica.

Remarks: Previously considered a synonym of *Mordilla kristenseni* it is here regarded as a valid species.

Pauleo jubatus Millen & Hamann 1992: 205-214, figs. 1-8.

Description: Background color yellow to pale orange with an opaque white midline from the head to the posterior end of the foot. Cerata showing a considerable color range depending on the color of the food. Rhinophores smooth.

Geographic Range: Belize, Honduras, Bahamas, Cayman Islands, Jamaica.

Natural History: Feeds on the gorgonian *Plexaurella dichotoma*. The cerata break off easily when disturbed.

Phidiana lynceus
Bergh, 1867

"Lynx phidiana"

Size: Up to 45 mm long

photos J. Hamann - left: typical specimen, Aruba; right: overgrown specimen, St. Lucia

photos A. Valdés & © L. Ianniello - left: Manzanillo, Costa Rica (AV); right: Florida (LI)

Learchis poica
Ev. Marcus & Er. Marcus, 1960

Size: Up to 25 mm long

photos J. Hamann & © T. Gosliner - left: dark cerata, Aruba (JH); right: light cerata, Cayman Islands (TG)

Phidiana lynceus Bergh 1867: 104-116, pl. 3, fig. B, pl. 4, fig. A.
Phidiana selencae Bergh 1879c: 560-563, pl. 6, figs. 10-18.
Phidiana brevicauda Engel 1925: 35-38, figs. 4-6.
Phidiana adiuncta Ortea, Caballer & Moro 2004: 86-94, fig. 3, pl. 1, figs. E-F.

Description: Background color translucent gray with a dorsal white or blue line that splits on the head and enters the oral tentacles. The line can be broad, narrow, or completely absent. Orange pigment on the oral tentacles and rhinophores. Larger cerata with white apices. Rhinophores annulate.

Geographic Range: Florida, Mexico, Costa Rica, Colombia, Venezuela, Bahamas, Aruba, Curaçao, Bonaire, Jamaica, Virgin Islands, Guadeloupe, Martinique, St. Maarten/St. Martin, St. Lucia, Barbados, St. Vincent & the Grenadines, Brazil.

Remarks: Ortea *et al.* (2004) differentiated species based on rhinophore and head morphology, however these characteristics show intraspecific variation (see rhinophore photos below). *Phidiana riosi* García & Troncoso, 2003 from Brazil appears to be a distinct and valid species.

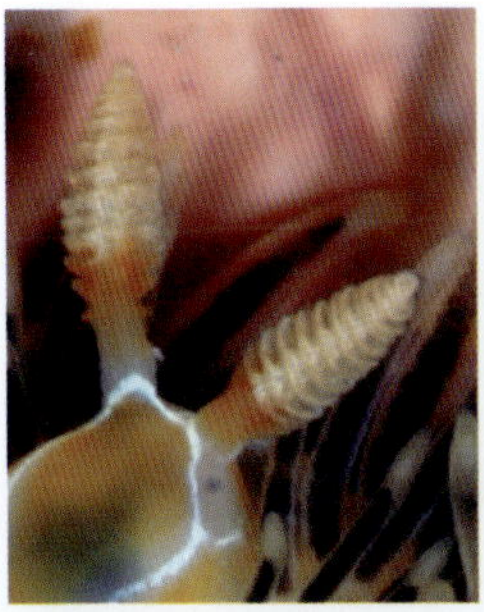

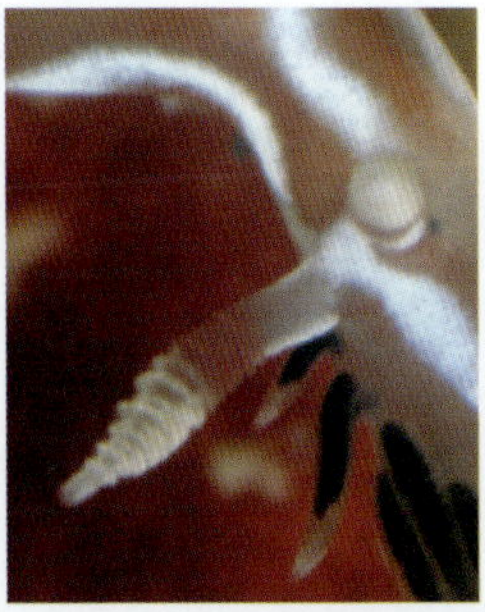

photos A. Valdés & J. Hamann - rhinophore variability; left: Costa Rica - AV (type locality of *Phidiana adiuncta*), specimen with equal lamellae; center left: Florida - AV, specimen with equal lamellae; center right: Aruba - JH, specimen with larger and smaller lamellae; right: Curaçao - AV, specimen with equal lamellae.

Learchis poica Ev. Marcus & Er. Marcus 1960a: 183, figs. 82-86.

Description: Background color translucent white to orange, with white patches on the head and behind the rhinophores. Cerata translucent clear with a few scattered opaque white spots and a subterminal white band. Digestive gland duct can range in color from light to dark brown. Rhinophores annulate.

Geographic Range: North Carolina, South Carolina, Florida, Mexico, Costa Rica, Bahamas, Aruba, Curaçao, Cayman Islands, Puerto Rico, Virgin Islands, Jamaica, Martinique, St. Lucia, Barbados, St. Vincent & the Grenadines, Grenada.

Natural History: Found on hydroids and attached to *Sargassum*.

photo J. Hamann - Grenada, reddish specimen

Learchis evelinae
Edmunds & Just, 1983

Size: Up to 14 mm long

photos J. Hamann - left: Point du Bout, Martinique; right: Lark Cay, Belize

Cratena cf. *peregrina*
(Gmelin, 1791)

Size: Up to 25 mm long

photo © L. Ruda - Florida

Cratena piutaensis
Ortea, Caballer & Espinosa, 2003

Size: Up to 8 mm long

photo J. Hamann - Halifax, Grenada

Learchis evelinae Edmunds & Just 1983: 195-197, figs. 5A, 7A-D, 8D.

Description: Background color translucent with opaque white pigment; upper three quarters of the oral tentacles opaque white. Rhinophores multicolored with a basal translucent band, an orange mid region, and an opaque white upper third. There is a characteristic orange line or patch on each side of the head from the base of the oral tentacle to the base of the rhinophores. Cerata translucent with reddish ramification of the digestive gland; some specimens with dark or white cnidosacs. Rhinophores smooth.

Geographic Range: Belize, Barbados, Martinique.

Doris peregrina Gmelin 1791: 3105.

Description: Background color translucent gray with two conspicuous, superficial orange patches in front of the rhinophores. Cerata translucent with bright red branches of the digestive gland, an opaque white cnidosac and some superficial opaque white spots. Rhinophores and oral tentacles with a translucent orange basal half and a opaque white upper half. Rhinophores smooth.

Geographic Range: Amphiatlantic; western Atlantic: Florida.

Remarks: This animal is very similar to specimens of the eastern Atlantic species *Cratena peregrina* and most likely belong to the same species.

Cratena piutaensis Ortea, Caballer & Espinosa 2003: 130-132, figs. 1A-D, 4A, pl. 1, fig. A.

Description: Background color translucent gray with numerous opaque white patches covering the entire body and oral tentacles. Two purple patches on the sides of the head and a red spot at the frontal end. Cerata with a dark brown base followed by a blue band, a red band a yellow band and a gray apex. Rhinophores fused together at the base, with a few yellow lamellae and a translucent gray base.

Geographic Range: Costa Rica, Grenada.

photo J. Hamann - St. Lucia

Facelinidae sp. 1

Size: Up to 18 mm long

photo J. Hamann - Cubagua, Venezuela

Facelinidae sp. 2

Size: Up to 21 mm long

photo J. Hamann - Chimana Grande, Venezuela

Facelinidae sp. 3

Size: Up to 6 mm long

Description: Background color pale brown with three longitudinal dorsal lines and two lateral lines, one on each side; lines black edged with white. Cerata with a brown base, a white central area and a translucent gray apex. Rhinophores and oral tentacles brown. Rhinophores with a few lamellae.

Geographic Range: St. Lucia.

Remarks: This appears to be an undescribed species.

Description: Background color translucent gray with opaque white spots on the dorsum and two orange areas in front of the rhinophores. Oral tentacles with a yellow proximal half and a bluish distal half. Cerata with a pale brown base followed by an orange band and a gray apex. Rhinophores fused together at the base, with several yellow lamellae and a translucent gray base.

Geographic Range: Venezuela.

Remarks: This appears to be an undescribed species.

Description: Background color translucent gray with irregular opaque white spots all over. Cerata, oral tentacles and rhinophores with the same pattern. Cerata with white branches of the digestive gland and the cnidosacs opaque white. Rhinophores smooth, the opaque white spots concentrated at the apex.

Geographic Range: Venezuela.

Remarks: This appears to be an undescribed species.

Family GLAUCIDAE Gray, 1826

Diagnosis: Body elongate, slender, widest just behind the head. Head short with very small, simple tentacles, widely separated. Rhinophores reduced, positioned low on the side of the head. Cerata arranged in several arch-like groups, some of them pedunculate. Cerata long, conical. Anus opens behind the inter-hepatic space (cleioproctic), renal pore posterior to the anus (abanal). Jaws bilobed with a denticulate masticatory border. Radular teeth with a prominent conical cusp normally flanked by sharp denticles.

Biology: The only two members of this family spend their lives floating upside down in open water, and maintain buoyancy by swallowing air. Glaucids exhibit an extremely effective camouflage system. The foot and undersides of the cerata are blue or blue and white, helping to camouflage them from predators above, such as sea birds; because they float upside down this is effectively their dorsal surface. The true dorsal surface, which faces down in the water, is silvery grey to camouflage them from fish looking up from below. Glaucids feed exclusively on species of *Physalia* (Portuguese Man O' War or Blue Bottle) and on other colonial floating cnidarians such as *Velella* (By-The-Wind-Sailor), and *Porpita* (Blue Button). Like many other aeolid nudibranchs, glaucids ingest the stinging cells (nematocysts) from their prey and retain them for their own defense. However, because of the large size and high toxicity of the venom in the stinging cells of *Physalia*, glaucids can cause a sharp, painful sting when touched. The sting of a glaucid is especially dangerous to children, elderly people, asthmatics and people with allergies as it can cause respiratory distress. Because glaucids are pelagic species being pulled by oceanic currents, the two species have broad distribution ranges throughout the world's oceans.

photo © K. Meyer - Panama

Glaucus atlanticus
Forster, 1777

"Silver slug"
"Blue dragon"
"Portuguese Man O' War slug"

Size: Up to 40 mm long

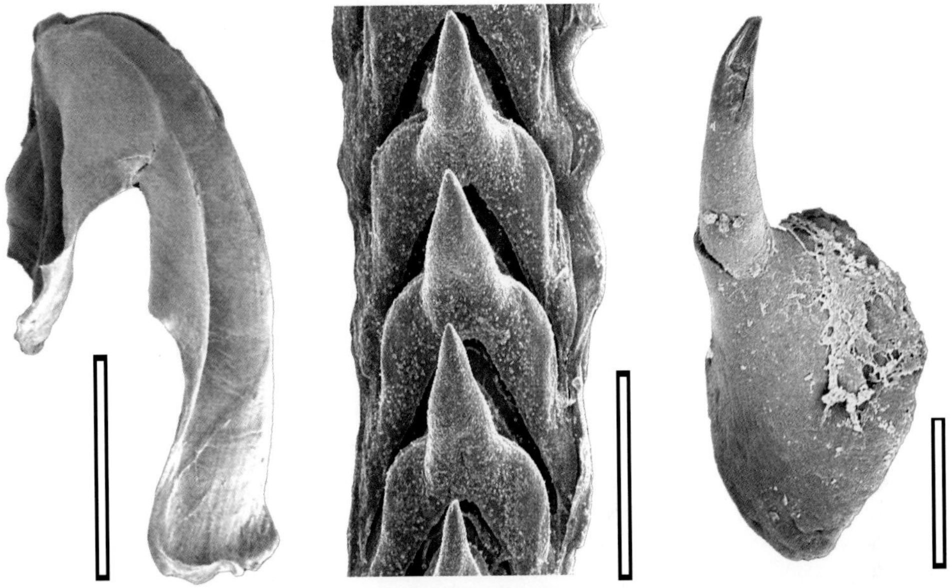

photos A. Valdés - SEM photos of *Glaucus atlanticus*; left: jaw, scale bar = 1 mm; center: radular teeth: jaw, scale bar = 100 µm; right: penial spine, scale bar = 100 µm

Glaucus atlanticus Forster 1777: 49.
[numerous other synonyms from this and many other geographic areas not listed]

Description: Background color silver, with the rhinophores, oral tentacles, and the lower edge of the cerata electric blue. Dorsum varies from dark blue to brown. First group of cerata pedunculate, with a short stalk; other two groups sessile. Head short and not distinctly separated from the body. Oral tentacles and rhinophores short and smooth.

Geographic Range: Cosmopolitan; Western Atlantic: Florida, Texas, Aruba, Bermuda, Jamaica, Barbados, Brazil.

Natural History: Feeds on colonial floating cnidarians such as *Physalia physalis*, *Velella velella* and *Porpita porpita*.

Family TERGIPEDIDAE Bergh, 1889

Diagnosis: Body elongate, with a swollen pericardial area. Foot usually wider than the body, rounded in front and often long and tapering posteriorly. Oral tentacles and rhinophores usually smooth. Cerata usually tubular, arranged in single oblique rows on each side of the body with more than one ceratal row in front of the pericardium. Anus high on the right side in the inter-hepatic space (acleioproct). Jaws more elongate than in other aeolid families, with the masticatory border denticulate, smooth or with bristles. Radula is uniseriate (only one tooth per row); teeth usually with a central cusp, either elevated or sunken, flanked by smaller denticles on each side.

Biology: Species of Tergipedidae are normally small aeolids, often found on subtidal rocky walls, on hydroid growths on man-made structures, and on the leaves of sea grasses and algae. Most species feed on hydroids, but members of the genus *Phestilla* (not recorded from the Caribbean), feed on hard corals

photo J. Hamann - Sharp Rocks, Bahamas

Cuthona pustulata
(Alder & Hancock, 1854)

Size: Up to 13 mm long

photo © L. Ianniello - Florida

Cuthona caerulea
(Montagu, 1804)

Size: Up to 25 mm long

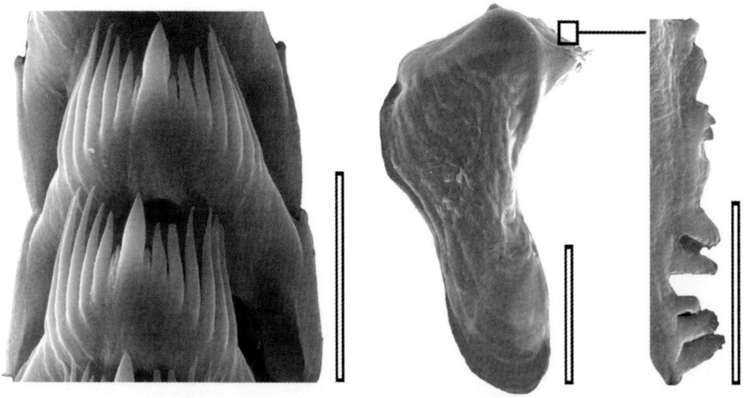

photos A. Valdés - SEM photos of *Cuthona perca*; left: radular teeth, scale bar = 30 µm; center: jaw, scale bar = 200 µm; right: detail of the jaw masticatory border, scale bar = 20 µm

Eolis pustulata Alder & Hancock 1854: 104.

Description: Background color translucent white, with a few scattered white dots on the dorsum. Cerata translucent with the branches of the digestive gland yellow and the cnidosacs opaque white. Rhinophores smooth, translucent white with some opaque white pigment near the apex.

Geographic Range: Amphiatlantic; Western Atlantic: Bahamas

Natural History: Found on floating *Sargassum*.

Remarks: This is the first record of this eastern Atlantic species in the western Atlantic.

Doris caerulea Montagu, 1804: 78, pl. 7, figs. 4-5.

Description: Background color translucent gray with a bright blue patch on the head. Oral tentacles pale gray-brown with conspicuous opaque white patches at the base. Rhinophores with a translucent base, a central white area and an orange-brown apical area. Cerata translucent white, with a blue band followed by a bright yellow band near the apex.

Geographic Range: Amphiatlantic; Western Atlantic: Florida, Brazil.

photo A. Valdés - Captiva Island, Florida

Cuthona perca
Er. Marcus, 1958

Size: Up to 15 mm long

photo © C. Redfern - Abaco, Bahamas

Cuthona tina
(Er. Marcus, 1957)

Size: Up to 4.5 mm long

photo © C. Redfern - Abaco, Bahamas

Cuthona barbadiana
Edmunds & Just, 1983

"Barbados cuthona"

Size: Up to 6 mm long

Cuthona perca Er. Marcus 1958: 45-52, figs. 81-87.

Description: Background color translucent gray with opaque white spots on the dorsum, rhinophores and oral tentacles similarly colored with a dark brown area in front of the rhinophores, and a few dark brown spots behind them. Cerata translucent with the branches of the digestive gland dark brown, an opaque white superficial ring near the apex and opaque white cnidosacs. Digestive gland visible in the posterior half of the body as a branched brown area entering the cerata.

Geographic Range: Cosmopolitan; Western Atlantic: Massachusetts (introduced), Florida, Colombia, Jamaica, Brazil.

Natural History: Found in intertidal lagoons or in harbor areas, sometimes in fouling communities.

Catriona tina Er. Marcus 1957: 459-462, figs. 187-198.

Description: Background color translucent grey with opaque white pigment and some brown spots all over the body. Oral tentacles translucent with some white pigment. Rhinophores opaque white with a dark brown ring in the middle. Cerata elongated, mostly opaque white, with the pale brown digestive branches visible in some areas.

Geographic Range: North Carolina, Florida, Costa Rica, Panama, Curaçao, Bonaire, Jamaica, St. Kitts, Barbados, Brazil.

Cuthona barbadiana Edmunds & Just 1983: 192, figs. 3E, 4D-F.

Description: Background color translucent gray, with visible orange-beige internal organs. Oral tentacles and rhinophores with irregular white lines at the base, a wine-red or purplish band near the middle, and a white upper half. A large white blotch on each side of the head and a regular pattern of white lines on the head and dorsum. Cerata with several longitudinal white lines and a purple area at the base.

Geographic Range: Barbados, Bahamas.

Natural History: Found on hydroids attached to *Sargassum.*

photo J. Hamann - St. Vincent

Cuthona iris
Edmunds & Just, 1983

Size: Up to 12 mm long

photo © M. Edmunds - Jamaica

Cuthona rubra
(Edmunds, 1964)

Size: Up to 4.5 mm long

photo J. Hamann - Guanaja, Honduras

Catriona maua
Ev. Marcus & Er. Marcus, 1960

Size: Up to 5 mm long

Cuthona iris Edmunds & Just 1983: 189-192, figs. 3C, 4A-C, 5B.

Description: Background color bright yellow with opaque white patches on the head, the cardiac area and forming a broad mid-line on the posterior end of the body; in some specimens all these are merged forming a single line. Oral tentacles uniform bright yellow. Rhinophores with a white base, a central orange area and a bright yellow apical area. Cerata bright yellow with a central bright blue area surrounded by opaque white pigment.

Geographic Range: Costa Rica, Barbados, St. Vincent & the Grenadines.

Selva rubra Edmunds 1964: 8-12, figs. 5B, 6A-D, 7A-C.

Description: Background color silvery gray with opaque white pigment on the distal half of the oral tentacles and rhinophores, on the head, and the sides of the body. Orange lines run from the base of the anterior cerata entering the rhinophores and the oral tentacles. Rhinophores with an orange ring. Cerata translucent with crimson branches of the digestive gland, fading towards the base; cnidosacs white.

Geographic Range: Jamaica.

Catriona maua Ev. Marcus & Er. Marcus 1960a: 177-180, figs. 74-79.

Description: Background color translucent greyish white with the dorsum covered with opaque white pigment, sometimes tinted with orange. A poorly defined reddish orange patch behind the eyes. Rhinophores and oral tentacles translucent gray with the upper halves pigmented with white and a reddish orange line on the lower third of the rhinophores. Cerata long, swollen, widest just above the mid-point; translucent white with a middle white band, followed by a translucent area and a second white band located just below the translucent tip. Ceratal base with an orange-red tinge formed by many small dots of dark red against the orange digestive system branches.

Geographic Range: Florida, Honduras, Mexico, Aruba, Curaçao, Bonaire, Jamaica, Barbados

Natural History: Found on hydroids attached to *Sargassum*.

Family AEOLIDIIDAE Gray, 1827

Diagnosis: Body elongate, broad or narrow, with the foot slightly wider or much wider than the rest of the body. Oral tentacles usually well developed; anterior corners of the foot angular or tentacular. Rhinophores simple and smooth, or with lamellae or papillae on the posterior side only or all around. Cerata often numerous and densely arranged, normally in rows, sometimes raised on a shared ridge. Cerata cylindrical and tapering or flattened. Anus opens among the cerata, behind the interhepatic space (cleioproctic). Radula with a single row of teeth, with a broad basal arch and numerous elongate and needle-shaped denticles.

photo J. Hamann - Bell Channel, Bahamas - *Spurilla sargassicola* feeding on anemones attached to floating *Sargassum*.

Biology: Most species feed on sea anemones and many of them sequester unicellular plant-like dinoflagellates (called zooxanthellae) from their prey. Zooxanthellae are common symbionts of cnidarians that are

photo J. Hamann - Miami, Florida

Spurilla neapolitana

(delle Chiaje, 1841)

"Neapolitan spurilla"

Size: Up to 40 mm long

left photo J. Hamann - Bell Channel, Bahamas; - comparison between *Spurilla neapolitana* (right) and *Spurilla sargassicola* (left) ingested by the aeolid while feeding on anemones. Instead of being digested with the cnidarian tissue, they are transported to specialized branches of the digestive gland. The zooxanthellae are maintained alive in the aeolid tissue and photosynthetic products produced by them are removed by the aeolid for its own nutrition. Species with zooxanthellae are usually recognizable by having brown reticulations in the body wall.

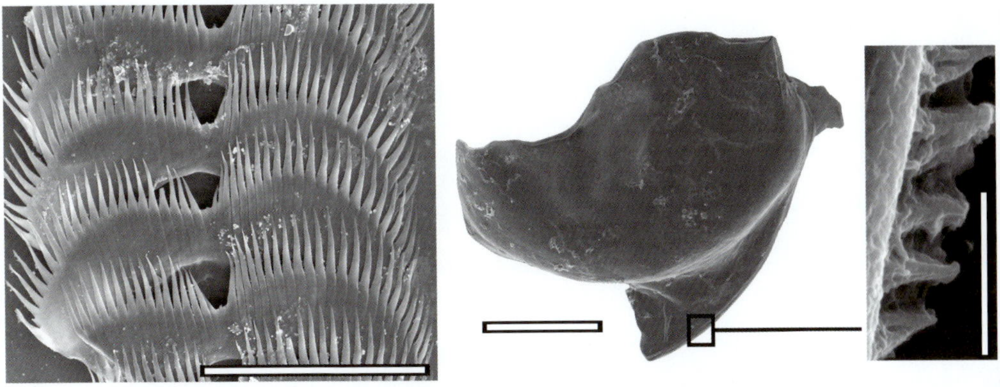

photos A. Valdés - SEM photos of *Aeolidiella stephanieae*; left: radular teeth, scale bar = 100 µm; center: jaw, scale bar = 500 µm; right: detail of the jaw masticatory border, scale bar = 3 µm

---

Eolidia neapolitana delle Chiaje 1841-44[1841]: 7, 11.
[several other synonyms from this and many other geographic areas not listed]

Description: Background color variable from pink to orange. Cerata translucent with the digestive gland ranging from gray to dark brown; some specimens have patches of opaque white pigment on the body, while others have none. Rhinophores lamellate; cerata flattened.

Geographic Range: Circumtropical and temperate seas, Western Atlantic: Florida, Texas, Belize, Honduras, Costa Rica, Colombia, Venezuela, Bahamas, Curaçao, Bermuda, Virgin Islands, Jamaica, Puerto Rico, Barbados, St. Vincent & the Grenadines, Brazil.

Natural History: Feeds on anemones of the genus *Aiptasia*.

photos A. Valdés - Left; adult specimen, Curaçao; right: juvenile, Sanibel Island, Florida

photo J. Hamann - Bell Channel, Bahamas

Spurilla sargassicola
(Kröyer, 1861)

Size: Up to 40 mm long

photo J. Hamann - Isla Borracha, Venezuela

Spurilla creutzbergi
(Er. Marcus & Ev. Marcus, 1970)

"Creutzberg's spurilla"

Size: Up to 30 mm long

photo J. Hamann - Isla Borracha, Venezuela

Aeolidiella indica
Bergh, 1888

Size: Up to 40 mm long

Facelina sargassicola Kröyer *in* Bergh 1861: 322.

Description: Background color light brown. Body and cerata covered by numerous opaque white spots and short dark brown lines. Rhinophores lamellate.

Geographic Range: Bahamas, Sargasso Sea.

Natural History: Found on *Sargassum*.

Remarks: This species is clearly different from *Spurilla neapolitana* and it is here regarded as a valid species.

Berghia creutzbergi Er. Marcus & Ev. Marcus 1970: 87-89, figs. 145-147.
Milleria ritmica Ortea, Caballer, & Espinosa 2003: 133-137, figs. 2, 4B, pl. 1, fig. B.

Description: Background color brownish, with dense white pigmentation on the dorsum and cerata. Rhinophores completely covered with large tubercles.

Geographic Range: Florida, Costa Rica, Venezuela, Curaçao, Cayman Islands, Barbados.

Natural History: The cerata rock from side to side when the animal is in motion, in a very distinctive rhythmic movement.

Remarks: The recently described new genus and species *Milleria ritmica* is a synonym.

Aeolidiella indica Bergh 1888: 781-783, pl. 78, figs. 1-2.
[several other synonyms from this and many other geographic areas not listed]

Description: Background color translucent white with rhomboidal opaque white areas on the medio-dorsal line, outlined with orange pigment, which extends from the oral tentacles and rhinophores to the posterior end of the body. Oral tentacles and rhinophores opaque white above, and orange below. Cerata with a subapical white band below the orange tip, each marked internally with a light brown branch of the digestive gland. Rhinophores smooth.

Geographic Range: Circumtropical; Western Atlantic: Venezuela, Curaçao, Puerto Rico, Brazil.

Natural History: Feeds on sea anemones.

Remarks: *Aeolidiella lurana* Er. Marcus & Er. Marcus is a synonym.

photo J. Hamann - South Padre Island, Texas

Aeolidiella benteva
(Er. Marcus, 1958)

Size: Up to 15 mm long

photo J. Hamann - Miami, Florida

Aeolidiella stephanieae
Valdés, 2005

"Stephanie's aeolidiella"

Size: Up to 20 mm long

photo J. Hamann - Piton Bay, St. Lucia

Berghia rissodominguezi
Muniain & Ortea, 1999

Size: Up to 52 mm long

Baeolidia benteva Er. Marcus 1958: 65-69, figs. 105-111.

Description: Background color translucent gray, with opaque white pigment on the center of the dorsum. Cerata grayish with an orange apex and subapical opaque white pigmentation. Rhinophores with large tubercles on the posterior side.

Geographic Range: Western Atlantic: Brazil, Texas.

Remarks: This species was originally described from Brazil as having large tubercles on the posterior side of the rhinophores and no data on the color of the living animals. The material from Texas illustrated here closely matches the anatomical descriptions of *Aeolidiella benteva* and therefore it is assigned to this species. Other records of *Aeolidiella benteva* are probably misidentifications.

Aeolidiella stephanieae Valdés 2005: 218-223, figs. 1-3.

Description: Background color translucent gray, with opaque white pigment on distal half of the rhinophores and oral tentacles. A few opaque white spots on the base of the oral tentacles, rhinophores, and sides of the body. Cerata grayish with diffuse white pigment on the base and an entirely opaque white apex, with a slight orange tinge in some specimens. Rhinophores with irregular folds.

Geographic Range: Florida.

Natural History: Feeds on sea anemones.

Remarks: This species has been misidentified as the Mediterranean species *Berghia coerulescens* (Laurillard, 1830). It is commercialized for aquarium control of *Aiptasia* infestations.

Berghia rissodominguezi Muniain & Ortea 1999: 144-150, figs. 1-2, pl. 1, fig. A.

Description: Background color translucent white, with oblique orange lines on the borders of the insertions of the cerata and in front of the rhinophores. Cerata translucent with reddish brown digestive diverticula and a white to yellow apex. Rhinophores and oral tentacles brownish red with white to yellow apices. Rhinophores tuberculate.

Geographic Range: Western Atlantic: Florida, Curaçao, Jamaica, St. Lucia, Brazil, Argentina.

Natural History: Found under rocks in intertidal areas.

Remarks: This species has been misidentified in the Caribbean as the Mediterranean species *Berghia coerulescens*. It could be a synonym of the eastern Atlantic *Berghia columbina* (García-Gómez & Thompson, 1990), but anatomical comparisons are necessary to confirm this point.

photo © L. Ianniello - Riviera Beach, Florida

Cerberilla tanna
Ev. Marcus & Er. Marcus, 1960

Size: Up to 25 mm long

photo © L. Ianniello - Riviera Beach, Florida

Cerberilla **sp.**

Size: Up to 25 mm long

photo J. Hamann - Caribo Cay, Belize

Baeolidia nodosa
(Haefelfinger & Stamm, 1958)

"Warty baeolidia"

Size: Up to 15 mm long

Cerberilla tanna Ev. Marcus & Er. Marcus 1960b: 259-260, figs. 18-19.

Description: Background color dirty yellow with pale brown dorsum and cerata. Cerata generally with a yellow base followed by triangular black spot and a bluish white apex. Rhinophores, oral tentacles and corners of the foot dirty bluish white. Rhinophores short, smooth.

Geographic Range: Florida, Texas.

Natural History: Inhabits sandy areas. Active at night.

Description: Background color yellow with grey dorsum and cerata. Cerata with a bright yellow patch. Rhinophores and oral tentacles navy blue. Head with a bright yellow and black pattern. Rhinophores long, smooth.

Geographic Range: Florida, St. Vincent & the Grenadines.

Natural History: Inhabits sandy areas.

Remarks: This appears to be an undescribed species.

Limenandra nodosa Haefelfinger & Stamm 1958: 418-423, fig. 1.

Description: Background color greenish brown, with white pigmentation on the dorsum. Dorsum with a longitudinal series of pink circles surrounded by broken circles of yellow spots. Rhinophores and oral tentacles greenish brown, the latter spotted with patches of white. White papillae on the rhinophores concentrated on the upper posterior half, with only a few smaller papillae elsewhere. Cerata covered with tubercles.

Geographic Range: Mediterranean and western Atlantic: Mexico, Belize, Honduras, Curaçao, St. Vincent & the Grenadines.

Natural History: Found on *Cladophora* and attached *Sargassum*.

REFERENCES

Abbott, R.T. 1949. A new Florida species of the tectibranch genus *Pleurobranchus*. *The Nautilus* 62: 73-78, pl. 5.

Abbott, R.T. 1952. Two new opisthobranch mollusks from the Gulf of Mexico belonging to the genera *Pleurobranchea* and *Polycera*. *Florida State University Studies* 7: 1-7.

Abraham, P.S. 1877. Revision of the anthobranchiate nudibranchiate Mollusca, with descriptions or notices of forty-one hitherto undescribed species. *Proceedings of the Zoological Society of London* for 1877: 196–269, pls. 27–30.

Adams, A. 1850. Monograph of the family Bullidae. *Thesaurus Conchyliorum* 2: 553-608, pls. 119-125.

Adams, A. 1855. Monographs of *Actaeon* and *Solidula*, two genera of gasteropodous Mollusca with descriptions of several new species from the Cumingian collection. *Proceedings of the Zoological Society of London* 22: 58-62.

Adams, C.B. 1840. Observations on some species of marine shells of Massachusetts, with descriptions of five new species. *Boston Journal of Natural History* 2: 262-288, pl. 4.

Adams, A. & Reeve, L. 1850. Mollusca, Part 3. *In:* Adams, A. ed., *The Zoology of the Voyage of H.M.S. Samarang, under the Command of Captain Sir Edward Belcher, C.B., F.R.A.S., F.G.S., During the Years 1843-1846*. Reeve, Benham & Reeve, London. Pp. 45-87, pls. 17-24.

Abraham, P.S. 1877. Revision of the anthobranchiate nudibranchiate Mollusca, with descriptions or notices of forty-one hitherto undescribed species. *Proceedings of the Zoological Society of London* [for 1877]: 196-269, pls. 27-30.

Alder, J. & Hancock, A. 1846. Notices of some new and rare British species of naked Mollusca. *Annals and Magazine of Natural History* 18: 289–294, pl. 4.

Alder, J. & Hancock, A. 1845-1855. *A Monograph of the British Nudibranchiate Mollusca: With Figures of all the Species*. Ray Society, London. Dates of publication: part 1, fam. 1 (pls 4-5, 26), fam. 3 (pls 3, 21, 24, 26, 34-36) [1845]; part 2, fam. 1 (pls 10, 13, 18, 23), fam. 3 (pls 1-2, 4, 6, 12, 15, 23, 30, 42) [1845]; part 3, fam. 1 (pls 6, 8, 19, 25), fam. 2 (pl. 3), fam 3 (pls 1a, 7, 19, 28, 31, 33) [1846]; part 4, fam. 1 (pls 7, 14, 20, 21, 24), fam. 2 (pl. 5), fam. 3 (pls 11, 13-14, 20, 25, 40) [1848]; part 5, fam. 1 (pls 1-2, 12, 15-16, 22), fam. 2 (pl. 4), fam. 3 (pls 5, 16-17, 27, 37, 39, 43) [1851]; part 6, fam. 1 (pls 3, 9, 11, 17), fam. 3 (pls 9-10, 18, 22, 29, 32, 41, 44) [1854]; part 7, fam. 1 (pls 21a, 27), fam. 2 (pls 1-2), fam. 3 (pls 38a, 45-48), appendix, pp. 1-54, i-xl [1855].

Alder, J. & Hancock, A. 1854. Notice of some new species of British Nudibranchiata. *Annals and Magazine of Natural History* (2)14: 102-105.

Alejandrino, A. & Valdés, A. 2006. Phylogeny and biogeography of the Atlantic and eastern Pacific *Hypselodoris* Stimpson, 1855 (Nudibranchia, Chromodorididae) with the description of a new species from the Caribbean Sea. *Journal of Molluscan Studies* 72: 189-198.

Baba, K. 1959. The genus *Petalifera* and a new species, *P. ramosa*, from Japan. *Publications of the Seto Marine Biological Laboratory* 7: 337-338.

Balch, F.N. 1899. List of marine Mollusca of Coldspring Harbor, Long Island, with descriptions of one new genus and two new species of nudibranchs. *Proceedings of the Boston Society of Natural History* 29: 133-162, pl. 1.

Barnard, K.H. 1933. Description of a new species of *Thecacera*. *Journal of Conchology* 19: 294-295.

Bergh, L.S.R. 1861. Om Forekomsten af Neldfiim hos Mollusker. *Videnskabelige Meddelelser fra Danske Naturhistorisk Forening* 1: 309–331, pl. 8.

Bergh, L.S.R. 1867. *Phidiana lynceus* og *Ismaila monstrosa*. *Videnskabelige Meddelelser fra Danske Naturhistorisk Forening* 7/9: 97-130, pl. 3, A-B, pl. 4, A-B.

Bergh, L.S.R. 1871. Beiträge zur Kenntniss der Mollusken des Sargassomeeres. *Verhandlungen der Kaiserlich-königlichen Zoologische-botanischen Gesellschaft* 21: 1273-1308, pls. 11-13.

Bergh, L.S.R. 1877. Malacologische Untersuchungen. *In*: Semper, C. ed., *Reisen im Archipel der Philippinen*, Theil 2, Heft 2. Kreidel, Wiesbaden. Pp. 429–546, pls. 54–61.

Bergh, L.S.R. 1879a. Die Doriopsen des atlantischen Meeres. *Jahrbucher der Deutschen Malakozoologischen Gesellschaft* 6: 42-64.

Bergh, L.S.R. 1879b. Neue Chromodoriden. *Malakozoologische Blatter* neue folge 1: 87-116, pl. 4.

Bergh, L.S.R. 1879c. Beiträge zur Kenntniss der Aeolidiaden, VI. Verhandlungen der Kaiserlich-königlichen Zoologische-botanischen Gesellschaft in Wien 28: 553-584, pls. 6-8.

Bergh, L.S.R. 1880a. Die Doriopsen des Mittelmeeres. *Jahrbucher der Deutschen Malakozoologischen Gesellschaft* 7: 297-328.

Bergh, L.S.R. 1880b. Malacologische Untersuchungen. *In*: Semper, C. ed., *Reisen im Archipel der Philippinen*, Theil 4, Heft 1. Kreidel, Wiesbaden. Pp. 1-78, pls. A-F.

Bergh, R. 1883. Beiträge zu einer Monographie der Polyceraden. *Verhandlungen der Kaiserlich-königlichen Zoologische-botanischen Gesellschaft* 33: 135-180, pls. 6-10.

Bergh, L.S.R. 1884. Malacologische Untersuchungen. *In*: Semper, C. ed., *Reisen im Archipel der Philippinen*, Theil 3, Heft 15. Kreidel, Wiesbaden. Pp. 647-754, pls. 69-76.

Bergh, L.S.R. 1888. Malacologische Untersuchungen. *In*: Semper, C. ed., *Reisen im Archipel der Philippinen*, Theil 3, Heft 16. Kreidel, Wiesbaden. Pp. 755-814, pls. 77-81.

Bergh, L.S.R. 1890. Report on the nudibranchs. Report on the results of dredging, under the supervision of Alexander Agassiz, in the Gulf of Mexico (1877-1878) and in the Caribbean (1879-1880), by the U.S. Coast Survey steamer "Blake," Lieut.-Commander C.D. Sigsbee, U.S.N., and Commander J.R. Barlett, U.S.N., commanding. *Bulletin of the Museum of Comparative Zoology* 19: 155-181, pls. 1-3.

Bergh, L.S.R. 1893. Die gruppe der Doridiiden. *Mittheilungen aus der Zoologischen Station zu Neapel Zugleich Repertorium für Mittelmeerkunde* 6: 107-135.

Bergh, L.S.R. 1894. XIII. Die Opisthobranchien. Reports on the dredging operations off the west coast of Central America to the Galapagos, to the west coast of Mexico, and in the Gulf of California, in charge of Alexander Agassiz, carried on by the U.S. Fish Commission steamer "Albatross", during 1891, Lieut. Commander Z.L. Tanner, U.S.N., commanding. *Bulletin of the Museum of Comparative Zoology* 25: 125-233, pls. 1-12

Bergh, L.S.R. 1897. Die Pleurobranchiden. Malacologische Untersuchungen. *In*: Semper, C. ed., *Reisen im Archipel der Philippinen*, Theil 4 Heft 1. Kreidel, Wiesbaden. Pp. i-v + 1-116, pls. 1-8.

Blainville, H. M. de. 1817. *Dictionnaire des Sciences Naturelles dans lequel on Traite Méthodiquement des Différens Êtres de la Nature, Considérés soit en Eux-Mêmes, d'Après l'État Actuel de nos Connoissances, soit Relativement àl'Utilité qu'en Peuvent Retirer la Médecine, l'Agriculture, le Commerce et les Arts*, Vol. 5, supplement. Levrault, F.G., Paris. Pp. 1-160.

Brandon, M. & Cutress, C.E. 1985. A new *Dondice* (Opisthobranchia: Favorinidae), predator of *Cassiopea* in southwest Puerto Rico. *Bulletin of Marine Science* 36: 139-144.

Bruguière, J.-G. 1792. Histoire Naturelle des Vers. *In: Encyclopédie Méthodique*, Vol. 1. Panckoucke, Paris. Pp. 345-757.

Bush, K.J. 1885. Additions to the shallow-water Mollusca of Cape Hatteras, N.C., dredged by the U.S. Fish Commission Steamer 'Albatross,' in 1883 and 1884. *Transactions of the Connecticut Academy of Arts and Sciences* 6: 453-480, pl. 45.

Caballer, M., Ortea, J. & Espinosa, J. 2001. Descripción de una nueva especie de *Eubranchus* Forbes, 1834. *Avicennia* supplement 4: 55-56, pl. 2.

Cantraine F.J. 1835. Diagnoses ou descriptions succinctes de quelques espèces nouvelles de mollusques. *Bulletin de l'Académie Royale des Sciences et Belles-lettres de Bruxelles* 2: 380-411.

Clark, K.B. 1977. *Phyllaplysia smaragda* (Opisthobranchia: Notarchidae), a new anaspidean from Florida. *Bulletin of Marine Science* 27: 651-657.

Clark, K.B. 1982. A new *Volvatella* (Mollusca: Ascoglossa) from Bermuda, with comments on the genus. *Bulletin of Marine Science* 32: 112-120.

References

Clark, K.B. & Goetzfried, A. 1976. *Lomanotus stauberi*, a new dendronotacean nudibranch from central Florida (Mollusca: Opisthobranchia). *Bulletin of Marine Science* 26: 474-478.

Conrad, T.A. 1846. Descriptions of new species of fossil and Recent shells and corals. *Proceedings of the Academy of Natural Sciences of Philadelphia* 3: 19-27, pl. 1.

Costa, A. 1866. Saggio sui molluschi eolididei del Golfo di Napoli. *Annuario del Museo Zoologico della Reale Università di Napoli* 3: 59-90.

Costa, A. 1867a. Saggio sui molluschi eolididei del Golfo di Napoli. *Annuario del Museo Zoologico della Reale Università di Napoli* 4: 26-37.

Costa, A. 1867b. Illustrazione di due generi di molluschi nudibranchi. *Rendiconto dell'Accademia delle Scienze Fisiche e Matematiche di Napoli* 6: 136-137.

Cuvier, G.L. 1817. *Le Règne Animal Distribué d'après son Organisation, pour Servir de Base à l'Histoire Naturelle des Animaux et d'Introduction à l'Anatomie Comparée*, Tome II Contenant les reptiles, les poissons, les mollusques et les annélides. M. le Cher. Chevalier, Paris. 550 pp.

Dall, W.H. 1980. A new species of *Tylodina*. *The Nautilus* 3: 121-122.

Dall, W.H. 1881. Reports on the results of dredging, under the supervision of Alexander Agassiz, in the Gulf of Mexico, and in the Caribbean Sea, 1877-79, by the United States Coast Survey Steamer 'Blake'. *Bulletin of the Museum of Comparative Zoology* 9: 33-144.

Delle Chiaje, S. 1841-44. *Descrizione e Notomia degli Animali Invertebrati della Sicilia Citeriore Osservati Vivi Negli Anni 1822-1830*. Stabilimento Tipografico di C. Batelli e Co., Napoli. Publication dates: vol. 1-4 [1841], pls. 1-173 [1841], pls. 174-181 [1844].

Deshayes, G.P. 1857. Note sur différents mollusques de la Guadeloupe, envoyés par M. Schramm. *Journal de Conchyliologie* 6: 137-143.

Dillwyn, L.W. 1817. *A Descriptive Catalogue of Recent Shells*. John & Arthur Arch, Cornhill. xii + 1092 [29] pp.

Eales, N.B. 1960. *Petalifera habei*, a new species from Japan. *Proceedings of the Malacological Society of London* 34: 12-18.

Edmunds, M. 1963. *Berthelinia caribbea* n. sp., a bivalved gastropod from the West Atlantic. *Journal of the Linnean Society, Zoology* 44: 731-739.

Edmunds, M. 1964. Eolid mollusca from Jamaica, with descriptions of two new genera and three new species. *Bulletin of Marine Science of the Gulf and Caribbean* 14: 1-32.

Edmunds, M. & Just, H. 1983. Eolid nudibranchiate Mollusca from Barbados. *Journal of Molluscan Studies* 49: 185-203.

Eliot, C.N.E. 1905. Nudibranchs from the Indo-Pacific, I. Notes on a collection dredged near Karachi and Maskat. *Journal of Conchology* 11: 237-256.

Eliot, C.N.E. 1908. Reports on the marine biology of the Sudanese Red Sea. XI. Notes on a collection of nudibranchs from the Red Sea. *Journal Linnean Society London, Zoology* 31: 86-122.

Engel, H. 1925. Westindische opisthobranchiate Mollusken. *Bijdragen tot de Dierkunde* 24: 33-80.

Engel, H. 1927. Westindische opisthobranchiate Mollusken II. *Bijdragen tot de Dierkunde* 25: 83-122.

Espinosa, J. & Ortea, J. 2004. Nuevas especies de moluscos gasterópodos marinos (Mollusca: Gastropoda) de las Bahamas, Cuba y el Mar Caribe de Costa Rica. *Revista de la Academia Canaria de Ciencias* 15: 207-216.

Fischer, P. 1857. Description d'espèces nouvelles. *Journal de Conchyliologie* 5: 273-277, pls. 8, 11.

Forster, J.G.A. 1777. *A Voyage Round the World, in His Britannic Majesty's Sloop, Resolution, Commanded by Capt. James Cook, During the Years 1772, 3, 4, and 5*, Vol. 1. White, London, 53 pp.

Gabb, W.M. 1873. On the topography and geology of Santo Domingo. *Transactions of the American Philosophical Society* new series 15: 49-259, 2 maps.

Gmelin, J.F. 1791. *Caroli a Linné Systema Naturae*, 13th Ed., Vol. 1, Pars 6. G.E. Beer, Lipsiae. Pp. 3021-3910.

Gosliner, T.M. 1989. Revision of the Gastropteridae (Opisthobranchia: Cephalaspidea) with descriptions of a new genus and six new species. *The Veliger* 32: 333-381.

Gosliner, T.M. 1994. New records of Flabellinidae (Opisthobranchia: Aeolidacea) from the tropical Americas, with descriptions of two new species. *Proceedings of the California Academy of Sciences* 48: 171-183.

Gosliner, T.M. & Armes, P.T. 1984. A new species of *Gastropteron* from Florida (Gastropoda: Opisthobranchia). *The Veliger* 27: 54-64.

Gosliner, T.M. & Ghiselin, M.T. 1987. A new species of *Tritonia* (Opisthobranchia: Gastropoda) from the Caribbean Sea. *Bulletin of Marine Science* 40: 428-436.

Gosliner, T.M. & Kuzirian, A.M. 1990. Two new species of Flabellinidae (Opisthobranchia: Aeolidacea) from Baja California. *Proceedings of the California Academy of Sciences* 47: 1-15.

Gould, A.A. 1852. Mollusca and shells. *In: United States Exploring Expedition during the years 1838, 1839, 1840, 1841, 1842. Under the Command of Charles Wilkes, U.S.N.*, Vol. 12. Gould & Lincoln, Boston, pp. i-xv + 1-510.

Gould, A.A. 1870. *Report on the Invertebrata of Massachusetts, comprising the Mollusca, Crustacea, Annelida, and Radiata*, 2nd Ed. W.G. Binney, Boston. v + 524 pp., pls. 16-27.

Gray, J.E. 1825. A list and description of some species of shells not taken notice of by Lamarck (continued). *Annals of Philosophy* (2)9: 407-415.

Haefelfinger, H.-R. & Stamm, R.A. 1959. *Limenandra nodosa* gen. et. spec. nov. (Nudibranch., Aeolidiidae propr.), un opisthobranche nouveau de la Méditerranée. *Vie et Milieu* 9: 418-423.

Hamann, J.C. & Farmer, W.M. 1988. Two new species of *Plocamopherus* from the western warm water Atlantic. *The Veliger* 31: 68-74.

Heilprin, A. 1888. Contributions to the natural history of the Bermuda Islands *Proceedings of the Academy of Natural Sciences of Philadelphia* [for 1888]: 302-328, pls. 14-16.

Ihering, H. von. 1886. Zur Kenntniss der Nudibranchien der brasilianischen Küste. *Jahrbücher der Deutschen Malakozoologischen Gesellschaft* 13: 223-240, pl. 9.

Ihering, H. von. 1915. Die Opisthobranchien der brasilianischen Kuste. *Nachrichtsblatt der Deutschen Malakozoologischen Gesellschaft* 47:133-143.

Jaume, M.L. 1945. El género *Oxynoe* en Cuba occidental. *Revista de la Sociedad Malacologica "Carlos de la Torre"* 3: 19-25, pl. 2.

Jensen, K.R. 1980. *Oxynoe azuropunctata*, n. sp., a new sacoglossan from the Florida Keys (Mollusca: Opisthobranchia). *Journal of Molluscan Studies* 46: 282-292.

Lea, H.C. 1845. A description of some new species of marine shells inhabiting the coast of the United States. *Proceedings of the Boston Society of Natural History* 1: 204-205.

Lightfoot, J. 1786. *A Catalogue of the Portland Museum, Lately the Property of the Duchess Dowager of Portland, Deceased: Which will be Sold by Auction, by Mr. Skinner & Co. on Monday the 24th of April 1786, and the Thirty-Seven Following Days at Twelve O'Clock*. Skinner, London. viii + 194 pp., 1 pl.

Linnaeus, C. 1758. *Systema Naturae*, 10th Ed., Vol. 1. Laurentii Salvii, Holmiae. 824 pp.

MacFarland, F.M. 1909. The opisthobranchiate Mollusca of the Branner-Agassiz expedition to Brazil. *Leland Stanford Junior University Publications* 2: 1-104, pls. 1-19.

Marcus, Er. 1955. Opisthobranchia from Brazil. *Boletim da Facultade de Filosofia, Ciências e Letras da Universidade de São Paulo, Zoologia* 20: 89-261, pls. 1-30.

Marcus, Er. 1957. On Opisthobranchia from Brazil (2). *Journal of the Linnean Society of London* 43: 390-486.

Marcus, Er. 1958. On western Atlantic opisthobranchiate gastropods. *American Museum Novitates* 1906: 1-82.

Marcus, Er. 1961. Opisthobranchia from North Carolina. *Journal of the Elisha Mitchell Sciences Society* 77: 141-151.

Marcus, Er. 1964. A new species of *Polycera* (Nudibranchia) from California. *The Nautilus* 77: 128-131.

References

Marcus, Er. & Marcus, Ev. 1970. Opisthobranchs from Curaçao and faunistically related regions. *Studies on the Fauna of Curaçao and other Caribbean Islands* 33: 1-129.

Marcus, Ev. 1970. On some opisthobranchs from Cananeia. *Boletim de Zoologia e Biologia Marinha* 27: 207-228.

Marcus, Ev. 1971. Opisthobranchs from northern Brazil. *Bulletin of Marine Science* 20: 922-951.

Marcus, Ev. 1972. On some opisthobranchs from Florida. *Bulletin of Marine Science* 22: 284-308.

Marcus, Ev. 1976a. Opisthobranchia von Santa Marta, Colombia. *Studies on Neotropical Fauna and Environment* 11: 119-150.

Marcus, Ev. 1976b. Marine euthyneuran gastropods from Brazil (3). *Studies on Neotropical Fauna and Environment* 11: 5-23.

Marcus, Ev. 1979. Campagne de la Calypso au large des côtes atlantiques de l'Amérique du Sud (1961-1962). 31, Mollusca Opisthobranchia. *Résultats Scientifiques des Campagnes de la Calypso* 11: 131-137.

Marcus, Ev. 1980. Review of western Atlantic Elysiidae (Opisthobranchia Ascoglossa) with a description of a new *Elysia* species. *Bulletin of Marine Science* 30: 54-79.

Marcus, Ev. 1983. The Western Atlantic Tritoniidae. *Boletim de Zoologia* 6: 117-214.

Marcus, Ev. 1984. The western Atlantic warm water Notaspidea (Gastropoda, Opithobranchia), Parte 2. *Boletim de Zoologia* 8: 43-76.

Marcus, Ev. & Gallagher, S.B. 1976. A new species of *Dendrodoris* from Florida. *The Veliger* 18: 353-356, pl. 1.

Marcus, Ev. & Hughes, H.P.I. 1974. Opisthobranch mollusks from Barbados. *Bulletin of Marine Science* 24: 498-532.

Marcus, Ev. & Marcus, Er. 1957. Sea-hares and side-gilled slugs from Brazil. *Boletim do Instituto Oceanografico* 6: 3-48.

Marcus, Ev. & Marcus, Er. 1960a. Opisthobranchs from American Atlantic warm waters. *Bulletin of Marine Science of the Gulf and Caribbean* 10: 129-203.

Marcus, Ev. & Marcus, Er. 1960b. Some opisthobranchs from the northwestern Gulf of Mexico. *Publications of the Institute of Marine Science, University Texas* 7: 251-261, pls. 1-3.

Marcus, Ev. & Marcus, Er. 1962. Opisthobranchs from Florida and the Virgin Islands. *Bulletin of Marine Science of the Gulf and Caribbean* 12: 450-488.

Marcus, Ev. & Marcus, Er. 1963. Opisthobranchs from the Lesser Antilles. *Studies on the Fauna of Curaçao and other Caribbean Islands* 19: 1-76.

Marcus, Ev. & Marcus, Er. 1966. The R/V Pillsbury deep-sea biological expedition to the Gulf of Guinea, 1964-65. 9.- Opisthobranchs from tropical west Africa. *Studies in Tropical Oceanography* 4: 152-208.

Marcus, Ev. & Marcus, Er. 1967a. American opisthobranch mollusks. *Studies in Tropical Oceanography* 6: 1-256, pl. 1.

Marcus, Ev. & Marcus, Er. 1967b. Opisthobranchs from the southwestern Caribbean Sea. *Bulletin of Marine Science* 17: 597-628.

Marcus, Ev. & Marcus, Er. 1967c. Some opisthobranchs from Sapelo Island, Georgia, U.S.A. *Malacologia* 6: 199-222.

Marcus, Ev. & Marcus, Er. 1968. *Flabellina engeli*, a new nudibranch from Curacao. *Beaufortia* 15: 139-142.

Marcus, Ev. & Marcus, Er. 1969. Euthyneure Meeresschnecken Brasiliens (2). *Beiträge zur Neotropischen Fauna* 6: 1-16.

Mattox, N.T. 1953. A new species of *Pleurobranchus* from the Caribbean (Tectibranchiata). *The Nautilus* 66: 109-114, pls. 9-10.

Maury, C.J. 1917. Santo Domingo type section and fossils. Part 1, Mollusca. *Bulletins of American Paleontology* 5: 165-415, pls. 1-39.

Menke, K.T. 1830. *Synopsis Methodica Molluscorum*. Author, Pyrmont. xvi + 169 pp.

Menke, K.T. 1853. Neue Arte der Gattung *Bulla*. *Zeitschrift für Malakozoologie* 10: 136-142.

Meyer, K.B. 1977. Dorid nudibranchs of the Caribbean coast of the Panama Canal Zone. *Bulletin of Marine Science of the Gulf and Caribbean* 27: 299-307.

Mikkelsen, P.S. and Mikkelsen, P.M. 1984. Comparison of *Acteocina canaliculata* (Say, 1826), *A. candei* (d'Orbigny, 1841), and *A. atrata* spec. nov. (Gastropoda: Cephalaspidea). *The Veliger* 27: 164-192.

Millen, S.V. & Hamann, J.C. 1992. A new genus and species of Facelinidae (Opisthobranchia, Aeolidacea) from the Caribbean Sea. *The Veliger* 35: 205-214.

Montagu, G. 1804. Description of several marine animals found on the south coast of Devonshire. *Transactions of the Linnean Society of London* 7: 61-85, pls. 6-7.

Montagu, G. 1815. An account of some new and rare British shells and animals. *Transactions of the Linnean Society of London* 11: 179-204.

Mörch, O.A.L. 1863. Contributions à la faune malacologique des Antilles danoises. *Journal de Conchyliologie* (3)11: 21-43.

Mörch, O.A.L. 1875. Synopsis molluscorum marinorum Indiarum Occidentalum. *Malakozoologische Blätter* 22: 142-184.

Muniain, C. & Ortea, J. 1999. First record of the genus *Berghia* Trinchese, 1877 (Opisthobranchia: Aeolidiidae from Argentina, with description of a new species. *Avicennia* 10/11: 143-150.

Nuttall, T.R. 1989. A new *Elysia* (Opisthobranchia: Ascoglossa) from the Florida Keys. *The Veliger* 32: 302-307.

Odhner, N.H.J. 1932. Beiträge zur Malakozoologie der Kanarischen Inseln. Lamellibranchien, Cephalopoden, Gastropoden. *Archiv för Zoologi* 23A: 1-116, pls. 1-2.

Odhner, N.H.J. 1936. Nudibranchia Dendronotacea - A revision of the system. *Memoires du Musée Royal d'Histoire Naturelle de Belgique* (2)3: 1057-1128, pl. 1.

Orbigny, A. d'. 1841. Mollusques. *In:* de la Sagra, R., ed., *Histoire Physique, Politique et Naturelle de l'île de Cuba*, Vol. 1. Arthus Bertrand, Paris. Pp. 1-240.

Ortea, J. 1979. Nota preliminar sobre *Rioselleolis anadoni* n. gen., n. sp. un nuevo Aeolidaceo (Mollusca: Opisthobranchia) capturado en Ribadesella Asturias, España. *Boletín de Ciencias de la Naturaleza del Instituto de Estudios Asturianos* 24: 131-141.

Ortea, J. 1981. Moluscos opistobranquios de las Islas Canarias. Primer parte: Ascoglosos. *Boletín del Instituto Español de Oceanografía* 6: 179-199.

Ortea, J. 1988. Moluscos opistobranquios del archipielago de Cabo Verde: C[h]romodorididae. *Publicaçoes Ocasionais da Sociedade Portuguesa de Malacologia* 11: 1-16, 1 pl.

Ortea, J. 2001a. El género *Doto* Oken, 1815 (Mollusca: Nudibranchia) en el mar Caribe: Historia natural y descripción de nuevas especies. *Avicennia* supplement 3: 1-46.

Ortea, J. 2001b. Descripción de una nueva especie de *Ancula* Loven, 1846. *Avicennia* supplement 4: 49, pl. 2.

Ortea, J. & Bouchet, P. 1983. Un nuevo Goniodorididae (Mollusca: Nudibranchiata) de las Islas Canarias. Vieraea 12: 49-54.

Ortea, J. & Caballer, M. 2002. Nuevos datos sobre el género *Eubranchus* Forbes, 1838 (Mollusca: Nudibranchia) en aguas templadas del Atlántico Oeste. *Avicennia* 15: 77-90.

Ortea, J. & Espinosa, J. 1996. Descripción de una nueva especie del genero *Elysia* Risso, 1818 (Opisthobranchia: Sacoglosa) recolectada en Puerto Morelos, México. *Avicennia* 4/5: 115-119.

Ortea, J. and Espinosa, J. 1998. Dos nuevas especies de moluscos marinos (Mollusca: Gastropoda) recolectadas en los subarchipiélagos Jardines del Rey y Jardines de la Reina, descritas en honor de los Reyes de España por su primera visita a Cuba. *Avicennia* 8/9: 1-5.

Ortea, J. & Espinosa, J. 2000a. Nueva especie del género *Janolus* Bergh, 1884 (Mollusca: Nudibranchia) de Cuba y Costa Rica. *Avicennia* 12/13 79-83.

Ortea, J. & Espinosa, J. 2000b. Nueva especie del género *Thuridilla* Bergh, 1872 (Mollusca: Sacoglossa) de Cuba y Costa Rica. *Avicennia* 12/13: 87-90.

References

Ortea, J. & Espinosa, J. 2001a. Descripción de una nueva especie de *Philinopsis* Pease, 1860 *Avicennia* supplement 4: 41, pls. 2-3.
Ortea, J. & Espinosa, J. 2001b. Descripción de una nueva especie de *Ercolania* Trinchese, 1872 *Avicennia* supplement 4: 45-47.
Ortea, J. & Espinosa, J. 2001c. Descripción de una nueva especie de *Dendrodoris* Ehrenberg, 1831. *Avicennia* supplement 4: 52-53, fig. 13, pl. 3, fig. E.
Ortea, J. & Espinosa, J. 2002. Nuevas especies del género *Elysia* Risso, 1818 (Mollusca: Sacoglossa) con caracteres singulares. *Avicennia* 15: 129-140.
Ortea, J., Espinosa, J. & Caballer, M. 2005. Nuevos taxones y registros de la familia Polyceridae (Mollusca: Nudibranchia) en las costas de Cuba. *Avicennia* 17: 101-106.
Ortea, J., Espinosa, J. & Camacho, Y. 1999. Especies del género *Polycera* Cuvier, 1816 (Mollusca: Nudibranchia) recolectadas en la epifauna de algas rojas del Caribe de Costa Rica y Cuba. *Avicennia* 10/11: 157-164.
Ortea, J., Caballer, M., & Espinosa, J. 2003. Nuevos aeolidaceos (Mollusca: Gastropoda) de Costa Rica. *Avicennia* 16: 129-142.
Ortea, J., Caballer, M. & Moro, L. 2001. Una nueva especie del género *Noumea* Risbec, 1928 (Mollusca: Nudibranchia) del Caribe arrecifal de Costa Rica, descrita con motivo del Quinto Centenario de su descubrimiento. *Avicennia* 14: 1-6.
Ortea, J., Caballer, M., & Moro, L. 2002. Descripción de los individuos jóvenes de *Hexabranchus morsomus* Marcus & Marcus, 1962 (Mollusca, Nudibranchia) con algunas consideraciones sobre la sistemática de la familia Hexabranchi[a]dae. *Revista de la Academia Canaria de Ciencias* 14: 193-200.
Ortea, J., Caballer, M. & Moro, L. 2004. Dos aeolidáceos con ceratas rojos de la región macaronésica y el mar Caribe (Mollusca: Nudibranchia). *Vieraea* 32: 83-96.
Ortea, J. & Llera, E.M. 1981. Un nuevo dórido (Mollusca: Nudibranchiata) de la isla Isabel, Nayarit, México. *Iberus* 1: 47-52.
Ortea, J., Luque, A.A., & Templado, J. 1990. Contributions to the knowledge of the genus *Aegires* Lovén, 1844 (Opisthobranchia: Doridoidea: Aegiretidae) in the North Atlantic, with descriptions of two new species. *Journal of Molluscan Studies* 56: 333-337.
Ortea, J. & Martínez, E. 1997. Una nueva especie de *Chelidonura* A. Adams, 1850 (Mollusca: Opisthobranchia: Cephalaspidea) de las costas de Cuba. *Avicennia* 6/7: 137-140.
Ortea, J., Moro, L. & Espinosa, J. 1997. El género *Doto* Oken, 1815 (Mollusca: Nudibranchia) en las islas Canarias y de Cabo Verde. *Avicennia* 6/7: 125-136.
Ortea, J., Valdés, A. & García-Gómez, J.C. 1996. Revisíon de las especies Atlánticas de la familia Chromodorididae (Mollusca: Nudibranchia) del grupo cromático azul. *Avicennia* supplement 1: 1-165.
Pease, W.H. 1871. Descriptions of nudibranchiate Mollusca, inhabiting Polynesia. *American Journal of Conchology* 6: 299-305, pls. 19-22.
Pelseener, P. 1892. Un nouveau nudibranche Méditerranéen. *Bulletin de la Société Malacologique de Belgique* 27: 19-21.
Petuch, E.J. 1987. *New Caribbean Molluscan Faunas.* Coastal Education & Research Foundation, Charlottesville, Virginia. v + 154 + A1-A4 pp., 29 pls.
Philippi, R.A. 1836. *Enumeratio Molluscorum Siciliae cum Viventium tum in Tellure Tertiaria Fossilium, quae in Itinere suo Observavit*, Vol. 1. Schropp, Berolini. xiv + 1-267 pp., 12 pls.
Pierce, S.K., Curtis, N.E., Massey, S.E., Bass, A.L., Karl, S.A. & Finney, C.M. 2006. A morphological and molecular comparison between *Elysia crispata* and a new species of kleptoplastic sacoglossan sea slug (Gastropoda: Opisthobranchia) from the Florida Keys, USA. *Molluscan Research* 26: 23-38.
Pola, M., Cervera, J.L. & Gosliner, T.M. 2005. A new species of *Tambja* (Nudibranchia: Polyceridae: Nembrothinae) from southern Brazil. *Journal of the Marine Biological Association of the United Kingdom* 85: 979-984.

Poorman, L.H. & Mulliner, D.K. 1981. A new species of *Crosslandia* (Nudibranchia: Dendronotacea) from the Gulf of California. *The Nautilus* 95: 96-99.

Quoy, J. & Gaimard, J. 1824. *Voyage autour du Monde, Entrepris par Ordre du Roi, sous le Ministère et Conformément aux Instructions de D. Exc. M. le Vicomte de Bouchage, Secrétaire d'État au Département de la Marine, Exécuté sur les Corvettes de S. M. "l'Uranie" et la "Physicienne", Pendant les Années 1817, 1818, 1819 et 1820, Zoologie*, Vol. 2. Aîné, Paris. 320 pp., 96 pls.

Quoy, J. & Gaimard, J. 1832-1833. *Voyage de Découvertes de "l'Astrolabe" Exécuté par Ordre du Roi, Pendant les Années 1826-1827-1828-1829, sous le Commandement de M.J. Dumont d'Urville*, Volume 2 & Atlas, Zoologie. Tastu, Paris. Dates of publication: 1-320 [1832], 321-686 [1833], pls. 1-26 [1833].

Rang, S. 1828. *Histoire Naturelle des Aplysiens. Première Famille de l'Ordre des Tectibranches.* Didot, Paris, 83 pp., 24 pls.

Rehder, H.A. 1939. New marine mollusks from the Western Atlantic. *The Nautilus* 53: 16-21, pl. 6.

Risbec, J. 1928. Contribution à l'étude des nudibranches néo-calédoniens. *Faune des Colonies Françaises* 2: 1-328, pls. 1-12.

Risso, A. 1818. Mémoire sur quelques Gastéropodes nouveaux, Nudibranches et Tectibranches observés dans la mer de Nice (1). *Journal de Physique, de Chimie, d'Histoire naturelle et des Arts* 87: 368-377.

Risso, A. 1826. *Histoire Naturelle des Principales Productions de l'Europe Méridionale et Particulièrement de Celles des Environs de Nice et des Alpes Maritimes*, Vol. 4. Levrault, Paris. 439 pp., 12 pls.

Rudman, W.B. 2002. *Hypselodoris marci* Marcus, 1970. *In*: *Sea Slug Forum*. Australian Museum, Sydney. Available at: http://www.seaslugforum.net/factsheet.cfm?base=hypsmarc

Russell, H.D. 1935. Some nudibranchs of Bermuda, with a description of a new species. *The Nautilus* 49: 59-61, pl. 4.

Say, T. 1826. Descriptions of marine shells recently discovered on the coast of the United States *Journal of the Academy of Natural Sciences of Philadelphia* 5: 207-221.

Schwengel, J.S. 1941. A genus and family of marine mollusks new to the United States. *The Nautilus* 55: 37-40, pl. 3.

Simroth, H. 1895. Opisthobranchia Gymnobranchia. Die Gastropoden der Plankton-Expedition. *Ergebnisse der Plankton-Expedition der Humboldt-Stiftung* 2: 1-206, pls. 1-22.

Smallwood, W.M. 1910. Notes on the hydroids and nudibranchs of Bermuda. *Proceedings of the Zoological Society of London* [for 1910]: 137-145.

Smith, E.A. 1872. A list of species of the genus *Planaxis*, with descriptions of eleven new species. *Annals and Magazine of Natural History* (4)9: 37-47.

Sordi, M. 1974. Nuova specie di mollusco nudibranchio del Mediterraneo. *Doriana* 5: 1-5.

Sowerby, G.B., II. 1868. Monograph of the genus *Haminea*. *Conchologia Iconica* 16: pls. 1-5.

Sowerby, G.B., II. 1869. Monograph of the genus *Atys*. *Conchologia Iconica* 17: pls. 1-5.

Swainson, W. 1840. *A Treatise on Malacology*. Longman, Orme, Brown, Green, & Longmans and John Taylor, London. viii + 419 pp.

Templado, J., Luque, A.A., & Ortea, J. 1987. A new species of *Aegires* Lovén, 1844 (Opisthobranchia: Doridacea: Aegiretidae) from the Caribbean Sea: *Aegires ortizi* spec. nov., with comparative descriptions of the North Atlantic species of this genus. *The Veliger* 29: 303-307.

Thiele, J. 1910. Molluskenfauna Westindiens. *Zoologischen Jahrbüchern* supplement 11: 109-132, pl. 9.

Thiele, J. 1931. *Handbuch der Systematischen Weichtierkunde*, Vol. 1. Fischer, Jena. vi + 778 pp.

Thompson, T.E. 1977. Jamaican opisthobranch molluscs I. *Journal of Molluscan Studies* 43: 93-140, pls. 1-3.

References

Thompson, T.E. 1980. Jamaican opisthobranch molluscs II. *Journal of Molluscan Studies* 46: 74-99.

Thompson, T.E., Cattaneo, R. and Wong, Y.M. 1990. Eastern Mediterranean Opisthobranchia: Dotidae (Dendronotoidea), Arminidae and Madrellidae (Arminoidea). *Journal of Molluscan Studies* 56: 393-413.

Trinchese, S. 1872. Un nuovo genere de la famiglia degli Eolididei. *Annali del Museo Civico di Storia Naturale di Genova* 2: 86-132, pls. 4-12.

Trinchese, S. 1881. Breve descrizione del nuovo genere *Lobiancoia*. *Rendiconto dell'Accademia delle Scienze Fisiche e Matematiche di Napoli* 20: 116.

Trinchese, S. 1891. Descrizione del nuovo genere *Bosellia*. *Memorie della Academia delle Scienze dell'Istituto di Bologna* (5)1: 773-778, 1 pl.

Usticke, G.W. 1959. *A Check List of Marine Shells of St. Croix*. Author, Christiansted, St. Croix. vi + 90 pp., 4 pls.

Valdés, A. 2004. Tropical western Atlantic species of *Diaulula* Bergh, 1878 (Mollusca, Nudibranchia), with the description of a new species. *Contributions in Science* 501: 1-7.

Valdés, A. 2005. A new species of *Aeolidiella* Bergh, 1867 (Mollusca, Nudibranchia: Aeolidiidae) from the Florida Keys, USA. *The Veliger* 47: 218-223.

Valdés, A. & Ortea, J. 1997. Review of the genus *Doriopsilla* Bergh, 1880 (Gastropoda: Nudibranchia) in the Atlantic Ocean. *The Veliger* 40: 240-254.

Valdés, A. & Ortea, J. 1998. A new species of *Doriopsilla* (Mollusca, Nudibranchia, Dendrodorididae) from Cuba. *Proceedings of the California Academy of Sciences* 50: 389-396.

Vanatta, E.G. 1901. New marine mollusks. *Proceedings of the Academy of Natural Sciences of Philadelphia* 53: 182-187, pl. 5.

Vayssière, A. 1897. Description de deux espèces nouvelles de Pleurobranchidés. *Journal de Conchyliologie* 44: 353-356.

Verrill, A.E. 1870. Contributions to zoology from the Museum of Yale College. No. 8. - Descriptions of some New England Nudibranchiata. *American Journal of Science and Arts* (2)50: 405-408.

Verrill, A.E. 1880a. Notice of recent additions to the marine Invertebrata, of the northeastern coast of America, with descriptions of new genera and species and critical remarks on others. *Proceedings of the United States National Museum* 3: 356-405.

Verrill, A.E. 1880b. Notice of the remarkable marine fauna occupying the outer banks off the southern coast of New England. *American Journal of Science and Arts* 3: 390-403.

Verrill, A.E. 1900. The nudibranchs and naked tectibranchs of the Bermudas. *Transactions of the Connecticut Academy of Arts and Sciences* 10: 545-550, pl. 66.

Verrill, A.E. 1901. Additions to the fauna of the Bermudas from the Yale Expedition of 1901, with notes on other species. *Transactions of the Connecticut Academy of Arts and Sciences* 11: 15-62, pls. 1-9.

Vogel, R.M. & Schultz, L.P. *Cargoa cupella*, new genus and new species of nudibranch from Chesapeake Bay and the generic status of *Okenia* Menke, *Idalia* Leuckart and *Idalla* Ørested. *The Veliger* 12: 388-393.

Weisbord, N.E. 1962. Late Cenozoic gastropods from northern Venezuela. *Bulletins of American Paleontology* 42: 1-672, 48 pls.

White, K.M. 1952. On a collection of molluscs from the Dry Tortugas, Florida. *Proceedings of the Malacological Society of London* 29: 106-120, pl. 6.

Woodring, W.P. 1928. Miocene Mollusks from Bowden, Jamaica, Part II. Gastropods and discussion of results. *Publications of the Carnegie Institute of Washington* 385: i-vii, 1-564, 40 pls.

GLOSSARY

aeolid – type of nudibranch possessing cerata rather than a plume-like gill on the back and no rhinophoral sheath.
acleioprotic – anus located dorsally, at mid-gut.
amphiatlantic – found on both (eastern and western) sides of the Atlantic Ocean.
annulate – referring to a type of rhinophore having a series of rings or bands along its stalk.
anterior – towards the head or front, situated in front.
anus – exit of intestine.
aperture – the opening or gape of a shell.
apex – the peak, or high point of a shell.
arborescent – many branches, tree like.
ascus – a unique radular structure of the Sacoglossa that retains worn teeth
bifurcate – branched, divided in two.
bilabiate – having two lips.
bipinnate – referring to the gill, branching twice from the central axis of the gill.
branchial plume – respiratory structure or gill, usually situated posteriorly on the dorsum.
buccal – mouth cavity.
buccal mass – the strong muscular portion of the digestive system posterior to the oral tube and anterior to the esophagus.
bulloid – having the appearance of a bubble, refers to a type of shell.
bursa – a pouch or sac.
carnivorous – eating animal tissue.
caryophyllidium (pl. caryophyllidia) – a specialized tubercle, surrounded by spicules, which has a sensory function.
cephalic – of or on the head.
cephalic shield – the fleshy plate that covers the anterior part of the body.
cephalic tentacles – the most anterior paired tentacles; head tentacles, oral tentacles.
ceras (pl. cerata) – finger-like respiratory structure sometimes bearing extensions of the digestive gland and occurring in groups of parallel series along dorsum.
chitin – a hard membranous material, made of hardened protein.
circumtropical – found in all tropical regions
clavate – gradually thickening near distal end.
clavus – the expanded or club shaped terminus of the rhinophore, above the stalk.
cleioproctic – anus located laterally, offset to the right side, near or among the cerata.
columella – inner margin of the aperture of a gastropod shell.
cnidosac – the sac at the tip of each ceras, where nematocysts are stored for defense.
cryptobranch – referring to the group of dorid nudibranchs which are capable of retracting their gill in a cavity.
cusp – apex, highest point.
denticle – a tooth on a radular teeth.
denticulate – having little teeth or raised points; with notches much finer than those referred to as serrate.
digitiform – finger-like.
distal – near the tip or end of a structure.
diverticulum – a pouch or structure, as a lobe of the digestive gland.
doridiform – shaped like a dorid nudibranch.

dorsum – the back or upper surface of an animal.
dorsal slit – in anaspideans, the dorsal opening between the right and left parapodia.
gill – a respiratory structure.
gizzard plate – chitinous plate in the digestive tract of some opisthobranchs.
hamate – a smooth hook shape.
head shield (see cephalic shield) – anterior fleshy portion of notum covering head in Cephalaspidea.
herbivorous – plant or algae eating.
hermaphrodite – an individual having both functional male and female sex organs.
hyponotal – underneath the notum, on the side of the foot.
inflated – swollen, usually referring to the cerata.
inter-hepatic space – space between branches of the liver.
involute – folded or rolled inward.
jaw – chitinous structure used to grasp prey while feeding.
labial – lips, around the mouth.
lamellate – referring to a type of rhinophore having a series of plates or lamellae.
lateral teeth – the teeth located to each side of the rachidian or central tooth on the radula.
maculations – spots.
mantle – in mollusks: the portion of the integumental fold which secretes the shell; reduced in primitive opisthobranchs, completely lost in adult nudibranchs.
mantle cavity – cavity housing gills and other organs in most gastropod mollusks including primitive opisthobranchs.
mantle glands – a series of defensive glands situated in the mantle of numerous dorid nudibranchs, particularly visible in chromodorids.
marginal teeth – radular teeth, usually along the margin of the radula. They differ in shape significantly from adjacent lateral teeth.
masticatory edge (or border) – cutting edge, usually of the jaw.
median, medial – toward the center or middle.
nematocyst – a microscopic stinging organelle produced by anemones, corals, and hydroids. Sea slugs can store them within a special sac, the cnidosac.
notum – the back surface of an opisthobranch, the dorsum.
nudibranch – a group of mollusks lacking a shell in the adult stage and having external respiratory appendages usually on the dorsal surface.
odontophore – a series of muscles and cartilage that support the radular teeth.
operculum – a plate serving to close the aperture when the animal retreats into its shell. Present in a few primitive opisthobranchs.
opisthobranch – mollusks having evolved from snail ancestors and possessing the following combination of characteristics: a reduced or completely absent shell, hermaphroditic genital apparatus, mantle cavity organs shifted to the right and posteriorly.
osphradium – folded sensory structure located in the gill cavity of some gastropod mollusks.
palmate – resembling a hand with fingers outstretched, referring to a type of cerata.
papilla (pl. papillae) – a finger-like extension with a basal diameter much smaller than its length and not containing digestive gland diverticulum.
papillate – covered with an abundance of finger-like extensions or papillae.
parapodia – flap-like lateral body extensions which arise ventrally.
perfoliate – referring to a rhinophore whose central shaft seems to pass through leaf-like extensions. Lamellate.